Ghillie James is an English cook and food writer. She was food editor of *Sainsbury's Magazine* for five years before she brought out her own cookery titles: *Jam, Jelly and Relish* (2010), *Fresh from the Freezer* (2011) and *Asia Light* (2014). She now lives in Singapore with her family.

GRAINS ARE GOOD

120 delicious ways to cook
with ancient grains

by Ghillie James
Photography by Jonathan Gregson

To Andrew, Wilbo, Mima and Helen.
We will eat potatoes again – I promise! xxx

First published in Great Britain in 2016 by
Kyle Books
192-198 Vauxhall Bridge Road
London, SW1V 1DX
www.kylebooks.co.uk

ISBN: 978 0 85783 370 9

10 9 8 7 6 5 4 3 2 1

Previously published in Great Britain as *Amazing Grains* in 2013

Photography: Jonathan Gregson
Design: Patrick Budge
Food styling: Annie Rigg
Prop Styling: Liz Belton
Project editor: Sophie Allen
Copy editor: Emily Hatchwell
Editorial Assistant: Tara O'Sullivan
Production: Gemma John and Nic Jones

A CIP record for this title is available from the British Library.

Colour reproduction by ALTA London
Printed and bound in China by 1010 Printing International Ltd

Grains everywhere

The world over, life begins with grains. Whether you are rich or poor, from Europe or Asia, Africa or Australasia, all young babies are weaned on rice. From this first simple bowl of goodness, our eyes are opened to the concept of eating and sharing food and to the wonders of grains. From beginnings that can be traced back to 12000 BC, and to the Middle East, grains are now eaten in every country of the world – in many countries as their staple food; nearly half the calories consumed around the world come directly from grains. And of course every country's favourite grain recipe is as important as the next – which has made filtering down the recipes in this book an incredibly tricky task.

Grains and 'pseudograins' (food products such as couscous and quinoa that are not technically grains but treated as such) are awesome ingredients, not only for their impressive nutritional content but also for their versatility. Simple, cheap and, on the whole, quick to cook, grains can be used as the star of the show or just a canvas to add to, in both sweet and savoury dishes. Transformed into everything from granola and porridge to stir-fries, salads, risottos and pilafs, they soak up dressings, absorb sauces and temper sharp or intense flavours. Grains are also filling – a few meagre ingredients can be transformed to create anything from a quick lunch to a tasty supper for a table of hungry guests.

Despite being eaten for thousands of years by ordinary folk all around the world, grains have not been immune to the vagaries of fashion. For a while, grains were banished along with other carbs as a casualty of the high-protein diets designed to help people shed pounds and achieve a waif-like body. Luckily, healthy, balanced eating now seems to be taking the helm, and health-giving grains are particularly in vogue. Heralded by everyone from Oprah to Jamie Oliver, specific grains and pseudograins are commonly described as superfoods.

With so many people being tested for wheat intolerances these days, many lesser known gluten-free grains such as buckwheat and teff are now also taking centre stage (though it was interesting to read some recent research suggesting that the number genuinely suffering from food intolerances is actually as low as 1–2 percent). For the majority, I believe that the quality of a person's diet is about what you include rather than what you exclude, and a good balance will benefit your body more than drastically cutting out any of the major food groups.

The range of grains out there is truly amazing – there are tens of thousands of varieties of rice, for example – and writing this book has been both an education and an adventure! Entering a market in Singapore, where I live, is like entering an Aladdin's cave of rice and other grains: there is so much choice and so much to learn. Singapore attracts a huge diversity of nationalities, bringing an amazing variety of culinary traditions to the city; on any trip to my small local supermarket I have my pick of about 50 varieties of rice and grains – Thai red rice, wholegrain Jasmine rice, black glutinous rice, teff, white cornmeal, jumbo couscous and Kamut, to name a few. And visiting the city's hawker stalls, the Arab quarter, the Indian quarter and Chinatown offers an education in the many and varied uses of these grains. This incredible country enables me to feast on a bulgar wheat salad in the morning, Thai pineapple rice for lunch and a biryani for supper (if I want, that is!).

What I have learnt is that cooking with some of the lesser known grains needn't be daunting. There are also so many ways to combine the textures and flavours of different grains in everyday cooking. A few storecupboard essentials and an easy cooking guide to follow will hopefully encourage you to diversify and experiment, avoiding crunchy or stodgy end results (or a cupboard full of grains that are starting to look past their prime!). Hopefully, you'll begin to realise the versatility of grains and their capacity to transform a nearly empty fridge into a tasty feast, packed full of nutrients and on the table in minutes.

What is a grain?

A grain is a seed or fruit of a plant hailing from the Poaceae (also known as Gramineae) family of grasses – and any of these grasses that produces an edible grain, such as barley, oats and wheat, is called a cereal. Cereals are annual plants, so they have only one growing season per year, and yield just one crop; the grains are harvested once the grasses are either dry, or dead. Some grains, such as rye, are winter grains, able to withstand cold, wet climates. Others, such as corn, are summer grains, which grow best in warm weather. The sheer variety of grains means that they are grown the world over.

Most grains are covered by a coarse, inedible husk (or hull), which surrounds the grain without being part of it. This needs to be removed by threshing so that the inner grain can be eaten. (The husk, once it has been removed, is known as the chaff and is used for animal fodder.) The inner grain left behind is made up of three edible parts:

Bran: This is the protective outer layer, which includes the very thin aleurone layer. It is highly nutritious, containing dietary fibre, protein and B vitamins. This part is removed when a grain is pearled.

Germ: This is the part of the seed which, if fertilised, will make a new plant. Although it accounts for just a tiny part of the grain, the germ is the powerhouse when it comes to nutrients. Sadly, it is removed when a grain is refined.

Endosperm: This is the largest part of the seed and gives the growing plant the food it needs to survive. It contains carbohydrate, protein and a small quantity of vitamins and minerals.

Different grain products

Different types of grains are suitable for different recipes. Whole grains may be the richest in fibre and some vitamins but I certainly don't use whole grains exclusively in cooking – pearl barley and white Basmati rice are just two examples of refined/processed grains that are delicious additions to recipes. Often, using a mixture of whole and pearled grains or pseudograins can be a great way to provide extra nutrients while keeping the recipe 'lighter', adding layers of different flavours, textures and colour to a recipe. I also like to combine grains with pulses, not only for their difference in taste and texture but also for the nutrients they provide – a recipe using wholegrain rice and puy lentils, for example, plus vegetables, perhaps some cheese and herbs would be about as close as you can get to nutritionally perfect. If you are like me and often eat only one proper main meal as well as breakfast, combining grains and pulses guarantees that you are ticking all the nutrient boxes and making sure you are having enough protein in your diet. This is particularly important for vegetarians.

Grains can be processed to varying degrees and in different ways, which influences the way they can be used in cooking:

Whole grain – an unrefined grain, which still has its fibre-rich bran, endosperm and germ intact. It can be used as is (as wheat berries or wholegrain rice, for example) or in processed form (such as rolled oats or wholegrain cracked wheat).

Polished/pearled/semi-pearled – a grain that has had some or all of its outer shell or bran and its germ removed, as in white Basmati or Thai Jasmine rice, Arborio rice and pearl barley, for example. These grains can then also be processed and used to make products such as plain couscous, flaked rice or white flour.

Cracked – grains that have been broken up, as in cracked wheat, producing a different texture and making them cook faster. In the case of bulgar wheat, the grains are parboiled before cracking, making them even quicker to cook.

Grits/groats/meal/semolina – grains that have been ground to varying degrees. Grits or groats are a midway point between cracked grains and flour. Meal tends to be like a very coarsely ground flour, though in the case of oatmeal the coarseness can vary. Semolina is coarsely ground endosperm, generally associated with wheat but also made from corn and rice.

Rolled/steel-cut/flaked – whole or pearled grains that are steamed, then rolled or cut into varying sizes and thicknesses. Examples include rolled oats, pinhead oatmeal, wheat flakes and Kamut flakes.

Roasted/toasted – grains that are roasted or parched to create various levels of flavour. Green wheat, for example, is roasted to produce a slightly smoky, rich-tasting grain called freekeh.

Flour – as fine as ground grains get. Flour made using whole grains results in a more bitty texture. Grains which have had the bran and germ removed (for plain flour, for example) can be ground to a very fine pure white powder. I decided not to include flour in the book due to the different treatment that it needs and the diversity of recipes that would need to be included.

Pseudograins

Pseudograins rank among the latest sexy, health-giving ingredients to be found on supermarket shopping lists and restaurant menus around the globe. Also known as pseudocereals or false grains, pseudograins (with the exception of couscous, which is a grain product) do not belong to the Poaceae family of cereal grasses; quinoa, chia, buckwheat and amaranth are the seeds of different species of broad-leafed plants. However, they are normally included with genuine grains, and have been included in this book because of their similar nutrient profile and the fact that they can be used in a very similar way to cereals; they often look like grains, too.

Grain production and its global impact

In ancient China, rice, wheat, barley and millet were among five sacred crops (the other being soybeans) revered by the people as the single most important things in life, ranking higher than pearls or jade. The ancient Chinese were not wrong. Cereal grains have been the principal component of the human diet around the world for thousands of years – to such an extent that rice, wheat, maize and, to a lesser degree, sorghum and millet, are now critical to the survival of billions of people. Associated with the world's dependence on grains, however, are some major economic, humanitarian, environmental and political issues.

That grains are used so widely around the world is due in large part to the fact that one or more cereals can be cultivated in each of the world's climates: the areas of major grain production stretch from the US and Canada, the European Union, Russia and Ukraine to India, South-east Asia, China and Australia. Compared with many other crops, grains are easy to cultivate, give a high yield, are easy to store and are very nutritious; and grain not used for human or animal consumption often forms the basis of biofuels. The United States is by far and away the biggest exporter of grain, supplying a quarter of the global grain trade.

Rice is the world's most important grain, being a staple food for nearly half of the world's population. It is a versatile crop that grows in many different climates and on all continents except Antarctica, flourishing as happily in the deserts of Saudi Arabia and on the coasts of Valencia as in the flooded rice plains of South-east Asia. However, more than 90 percent of rice is grown and consumed in Asia: China, India and Indonesia are the world's three largest producers, and Vietnam, Thailand, Burma and the Philippines all appear within the top ten. History has shown that there is a clear link between an expansion in rice cultivation and a

rapid rise in population growth, due largely to the fact that rice can support more people per unit of land than the other two principal staple grains, wheat and maize. To many Asians, rice is of truly fundamental significance, not just to their diet but to their entire lives. In several Asian languages the word for rice is the same as for 'meal' or 'food', and instead of, 'How are you?' the greeting 'Have you eaten rice today?' is used! Rice plays a part in many rituals and ceremonies and is viewed with great reverence, as a symbol of prosperity, a source of happiness and a cause for celebration.

From its origins in the Middle East where, fittingly, it remains the primary staple food, wheat now covers more of the earth's surface than any other commercial crop, and world trade in wheat is greater than for all other crops combined. Though it ranks second to rice in terms of its importance as a human food crop, wheat forms an increasingly large part of the diet for around one third of the world's population. Globally, wheat production has increased largely as a result of urbanisation and shifting tastes and preferences. While wheat has for centuries been a staple food in Europe, the Middle East and North Africa, it is growing in favour among countries such as China and Japan that have not traditionally consumed much of the stuff. China is now the world's main producer of wheat, followed by India (which has long been a major consumer of wheat: there are around 50 different types of Indian bread made from wholewheat flour). Wheat is also used for animal feed and industrial purposes, particularly when harvests are ravaged by rain and the grain becomes unsuitable for human consumption. In fact, wheat production cannot keep pace with global demand and population growth and, if this trend continues, it may become increasingly difficult to maintain a wheat supply for generations in the future.

Maize is the third most important food crop in the world, adopted as a staple or supplement in virtually every world region, and remains the primary staple for much of Latin America. Today, maize is a demonstration of globalisation and multinational commerce, with the States the world's pre-eminent corn grower and exporter, supplying more than 40 percent of the world's corn. If you think that maize is only found in corn-on-the-cob and Cornflakes, you may be surprised to learn that it appears in a baffling variety of foods and products. Carry out an ingredients label check the next time you're in the supermarket and see how often you come across glucose syrup (a.k.a. corn syrup).

It wouldn't be going against the grain to say that the demand for most types of grain is growing. Our long-suffering earth is under strain from a rising and more demanding population, clamouring for increasing amounts of grain, either as food or animal feed, not to mention fuel. Indeed, so ambitious are the biofuel quotas imposed in America and Europe that there are serious concerns about the impact on world food supplies (particularly of corn, the main source of biofuel). While in some countries the increased demand can be met domestically by improving yields, in many it cannot. There is little arable land that remains unfarmed and with a burgeoning population, farmers face a monumental challenge to produce the level of grain needed to feed the world's people.

Many of those who are dependent on rice for food live in increasingly populous developing countries; how can the earth's 112 rice-producing nations keep pace? Even China has started to import rice in recent years, which could herald a major and enduring shift in the global rice market. Meanwhile, in the Arab countries of the Middle East and North Africa, whose peoples comprise only 5 percent of the world's population, they import more than 20 percent of the world's grain: the scarcity of water and arable land means that domestic grain production cannot begin to cater for the needs of local people.

With water shortages spreading and rising global temperatures bringing more unpredictable weather, environmental factors are of major concern when it comes to grain production. Freak weather in some of the world's major food-producing regions in recent years has had a hugely negative impact worldwide, devastating crops and triggering food crises. From droughts in North and South America to flooding in Russia and Thailand, the impact of ruined grain harvests is severe and widespread, the inevitable consequence being soaring prices, which lead to higher prices of food as a whole. These wild price rises hit the poorest countries of the world hardest. Westerners spend around 15 percent of their income on food, but those living in developing countries spend around 75 percent, meaning that any change in food prices has a particularly dramatic impact on household budgets, raising the spectre of economic, social and political unrest. According to the US State Department, more than 60 food riots erupted worldwide between 2007 and 2009 in response to global costs reaching what were then all-time highs. Rising food prices have also been cited as a contributing factor to the 2011 Arab Spring revolution in Egypt.

The challenge is to produce a higher level of grain reserves in order to counteract harvest failures and subsequent shortages. The subject of grains is likely to remain one of the most hotly debated global topics of the 21st century. It will be interesting to see what the coming years hold.

A history of grain production

The fascinating story of how grains came to be a staple commodity for many of us is an ancient one. Around 10,000 years ago, the first Agricultural, or Neolithic, revolution began in an area known as the Fertile Crescent, which corresponds roughly to modern-day Israel, Lebanon, Jordan, Syria, Iraq and northern Egypt. Here began the key transition from a nomadic hunter-gatherer society to a more sedentary agricultural one. Grain laid the foundation on which many self-sustaining and self-supporting communities were built and marked the beginnings of a fundamental change in the development and evolution of civilisation – a process which spread across the world over the following eight millennia.

'It is probable that man learned to pound grain even before he could grow it.' Thus wrote Tom Stobart in his *Cook's Encyclopedia*. Before farming, humans survived by hunting animals and gathering wild plants. Gradually, the nature of the traditional food supply changed, with the discovery that wild grains could offer hearty sustenance. Man began to gather grains for food, nomadic herders became permanent settlers and the seeds of the plants were scattered to grow food. They would stash the grain away, before later sowing the seeds in special plots. At the Gilgal site near Jericho in the West Bank, excavators found some 260,000 grains of wild barley and 120,000 grains of wild oats stored in a stone building which dates back to 11,000 years ago, thus providing evidence of cultivation during the Neolithic Period. Over a period of thousands of years, early farmers shifted from the simple cultivation and harvesting of wild grains, to domesticating grains. By 8000 BC, farmers were producing more food than they needed and could swap grains and other foods for practical or decorative goods with artisans and traders. Many of the workers who built Egypt's pyramids at Giza were often paid in bread and beer.

In the ancient world, different grains became associated with particular areas – maize with ancient Mexico, rice with China and wheat and barley with the Near East. Primitive transportation technology meant that the trade in grains was limited, but this changed when the wheel was invented. As technology for both storing and transporting grains improved, grains began to move into previously unchartered territories. The unification of China and the pacification of the Mediterranean basin by the Roman Empire produced vast regional commodities markets at either end of Eurasia. Once the Roman system collapsed, feudalism predominated and farmers were driven to subsistence farming, making trade a thing of the past. However, when the Europeans started cultivating swathes of land in Russia, Australia and the Americas from the 15th century onwards, the grain trade really began to boom. Farmland in Britain and Eastern Europe was consolidated, the railway was born and the steam ship took trade from a local to an international level. As increasing numbers of trade routes opened up and as people began to emigrate, new varieties of grains spread across the world.

A chronology of grains
Due to sometimes conflicting evidence, the dates and facts listed in this chart are approximations!

12000BC – Einkorn, a species of wheat, is the first grain to be domesticated in the Fertile Crescent

10000BC – Emmer evolves from einkorn and emerges in Egypt as the second domesticated grain – Wild rice is found in North America and Asia – Bread, in the form of unleavened flatbread, is made for the first time

9300BC – Evidence of systematic storage of wild grains found near the Dead Sea in Jordan dates to this period

8500BC – Barley is domesticated in Mesopotamia (modern-day Iraq)

8000BC – The first domesticated rice is developed, most likely in the lower and middle Yangtse River Valley in China

6700BC – Maize is first domesticated in Mexico

6500BC – Cultivated emmer wheat reaches India and parts of Europe

5500BC – Millstones start being used for grinding flour

5000BC – Millet is domesticated and cultivated simultaneously in Asia and Africa – Cultivated emmer wheat reaches Spain – Barley is grown in Japan

4000BC – Quinoa is cultivated by the Incas in Peru, Bolivia and Chile

3000BC – Rye is domesticated in parts of Turkey, Armenia and Iran – The Egyptians are the first to produce leavened bread using yeast – Cultivated emmer wheat reaches England and Scandinavia – Sorghum is domesticated in sub-Saharan Africa

2600BC – Chia is domesticated in Mexico

2500BC – Rice is domesticated in India

2300BC – Freekeh is discovered in the Eastern Mediterranean

2000BC – Evidence dates oat grains to this period, when the Ancient Egyptian 12th Dynasty rules – The first-known noodles are made in China using millet

1500BC – Rice is domesticated in Africa

1000BC – Oats are domesticated in Europe – Buckwheat is cultivated in China

500BC – Spelt is commonly grown in England – Rice is grown over widely scattered areas of Asia, stretching from India to the Philippines

200–150BC – First evidence of couscous among the Berber people of North Africa

300–200BC – Rice is grown in the Middle East and Japan

AD 500–900 – Earliest reputed windmill for grinding grains designed in Persia

1400–1600 – Crop rotation is introduced

1500s – Colonists introduce grains to the New World, e.g. the Spanish introduce wheat and barley to South America

1600s – English and Dutch settlers bring barley to the United States

1700–1800 – The Industrial Revolution leads to advanced farming techniques, larger scale farming and improved quality of grain, and breadmaking becomes firmly established as a business and a trade

1800–1900 – The first refined flour is made to improve storage and satisfy demand

1900 onwards – Advances in crop breeding, mechanisation and management lead to better quality and yield of grains and more efficient production

Cultivation versus domestication

It is important to understand the difference between the cultivation and domestication of grains, as in chronological terms there can be thousands of years between the two. A simple explanation is that cultivated grains are simply wild grains that have been cared for, watered and harvested by a farmer. The process of plant domestication, on the other hand, involves altering the plant's native characteristics and genetic makeup, to the point where it cannot grow and reproduce without human intervention. A plant's most desirable characteristics are identified and grains which give the best yields are harvested selectively, with only the best seeds from previous crops being used.

Ethical issues in grain production

Due to the importance of grains as a commodity, both for food and fuel worldwide, there are unsurprisingly many controversial issues associated with grain production. Genetic modification and biopiracy are two of the main issues being debated.

As you can see from the Chronology, farmers were modifying crops long before the advent of genetics and 'modern' biotechnology. They selected plants to grow, first choosing from wild species and then cultivating and domesticating them (essentially changing their genetics), in order to produce the biggest and the best, resistant to disease and able to grow in varying temperatures and soils. Genetic modification, however, refers to a much bigger and more controversial subject, which has raised much public concern and debate. Genetically modified (GM) crops are those plants that have had their DNA modified by genetic engineering techniques. In most cases, this technique introduces a new trait to the plant, which wouldn't occur naturally, for example resistance to weedkilling herbicides or even drought.

GM foods have had mixed support worldwide. In the US, genetic modification has expanded into almost every area of food production. In many cases, the GM products are either added to animal fodder or sold as commodities which go into processed food. There is far less support to date among EU consumers, the majority of whom are still opposed to the development of genetically modified crops. Britain resoundingly rejected the first generation of commercial GM crops when they were first introduced in the 1990s. However, research and development continues. A wheat crop sown in the Hertfordshire countryside was genetically engineered to emit a repellent-smelling substance against insect pests. Nicknamed 'whiffy' wheat, it could combat aphid attacks that can cause upwards of £120 million worth of damage each year to the UK's most important cereal crop.

Another hot, ethical topic is biopiracy, the practice of claiming patents on a biochemical or genetic material to restrict its future use, while failing to pay fair compensation to the community from which it originates. So, for example, a developed country such as the US may be allowed to 'tweak' the germ plasm (the part of the germ that contains the genes) from a Jasmine rice seed sample collected in Thailand and then patent it as its own – the implication being that the corporation that owns the patent has exclusive rights to the production and distribution of the seeds. There are numerous examples of this worldwide and many ongoing cases being fought.

The popularity of biofuel production is also a hugely debated issue. Some argue that the energy used in production is actually more than that created, and others have concerns that the increasing use of grains to produce the fuels will have the potential to exacerbate hunger worldwide. Meat and biofuel production are jointly raising the cost of grain, adversely affecting the poorest people in the world. There is an obvious political, economical and environmental counter-argument to this, as the production of biofuels reduces both the dependence on hydrocarbons and fossil fuels and the amount of atmospheric carbon being produced.

Grains for health

According to the Department of Health in the UK, 33 percent of our food consumption should come from bread, rice, potatoes, pasta and other starchy food, of which as much as possible should be whole grain. However, the average household is eating far less than this. According to the findings, in order to be in tip-top health, we need to lower the amount of dairy, fat and sugary foods and drastically up the quantity of starchy foods and whole grains we are eating by around 70 percent. A variety of grains are becoming increasingly easy to find in our supermarkets. There really is no excuse not to use them, especially as they are a cheaper and healthier alternative to processed foods and ready-made meals, giving you a more balanced and wholesome diet.

The health benefits of eating grains
by Anna Lynch
BA(Hons), DipION(Nutr), ATMS

Whole grains play a valuable role in a balanced diet, providing carbohydrate for energy, fibre for a healthy digestive system and a range of important nutrients for health and wellbeing. 'But shouldn't we be avoiding grains?' I am often asked. 'Aren't they fattening?' The answer to this question is that we should try and avoid being overly dependent on just one grain (usually wheat), and we should certainly be choosing whole grains over their refined counterparts.

Here's an example. Client A has a boxed breakfast cereal every morning, followed by a couple of biscuits with his coffee mid-morning and a panini for lunch. He usually grabs a muffin during the afternoon and enjoys a pasta dinner. This client has effectively eaten refined wheat flour throughout the day, which offers little in the way of nutrition and fibre, yet plenty of quick-releasing carbohydrates that will raise blood glucose levels and be stored as fat if not burned as energy.

Client B also includes grains in his diet but the impact on his health is quite different. For breakfast he enjoys porridge made from whole rolled oats with chia seeds. Lunch is a bulgar wheat and lentil salad and dinner usually includes quinoa or brown rice. He likes to snack on home-made popcorn. Unlike Client A, Client B is including at least four different nutritious whole grains in his diet every day. As a result, he is boosting his nutrient and fibre intake and reducing his risk of colon cancer, type 2 diabetes, cardiovascular disease and obesity.

A grain is 'whole' when the entire grain seed is retained, that is the bran, germ and endosperm. Most of the goodness of the grain lies in the bran and the germ, including fibre, B vitamins, minerals and antioxidants. Modern food refining methods remove the bran and germ of the grain, leaving just the endosperm. This provides energy but most of the valuable nutrients have been lost. Foods made from refined grains include many breakfast cereals, snack foods, breads, pizza and biscuits. The benefits of including a variety of whole grains in our diets are numerous. Grains are a source of fibre, which helps to prevent constipation and provides a healthy environment for beneficial bacteria in the digestive system. A high-fibre diet has been associated with a decreased risk of developing colorectal cancer. A 2008 review of seven studies involving more than 285,000 people found that consuming an average of 2.5 servings of whole grains each day can lower the risk of cardiovascular disease by 21 percent, compared to a very low intake of whole grains. Further research has supported the role of whole grains in weight loss, reducing type 2 diabetes, and lowering cholesterol. In particular, it would appear that whole grains play a role in reducing inflammation in the body, which is believed to be a key driver in many chronic diseases.

Pseudograins are widely regarded as superfoods, and for good reason. They have a beneficial alkalising effect on the body, are gluten-free and contain a valuable range of nutrients including B vitamins, calcium, magnesium and amino acids. We are lucky enough today to have exciting grains available to us from all over the world. I hope the recipes in this book will inspire you to discover more about the wonderful diversity, texture and taste of grains and to experience the many health benefits they offer.

The Benefits of Low-GI Foods

When you eat a meal consisting of large quantities of certain carbohydrates, you can often feel very full and then, soon afterwards, very hungry. This is because some foods, such as certain breakfast cereals, short-grained white rice and processed white bread have high-GI (glycemic index) levels – meaning they are very quickly broken down by the body, providing a fast zap of energy. This is ideal for athletes needing a quick fix of energy to burn, but not for those of us who want to eat healthily, stay full for longer and avoid snacking in between meals. Therefore, while it's fine to include some high-GI foods in your diet, it's healthier to include as many low-GI foods as possible (or at least pair high- and low-GI foods). Here's a list of some low and high GI-foods containing grains:

Low- to medium-GI grains (including processed)

White and brown long-grain rice
Quinoa
Buckwheat
Rye
Pearl barley
Cracked wheat
Bulgar wheat
Pearled spelt
Rolled oats (not instant)
Multigrain/seeded bread
Muesli

High-GI grains (including processed)

Short-grained white rice
Glutinous rice
Instant white rice
White pasta
Millet
Puffed wheat
White bread
Cornflakes
Weetabix
Rice Krispies

Six quick ways to improve your diet using grains

1. Cook a combination of grains – brown, wild and white rice, quinoa with buckwheat, amaranth with couscous, etc. – rather than one on its own. Mixing up grains changes the taste and texture and adds loads of extra nutrients.
2. Sprinkle grains into a soup or broth, either keeping it chunky or blitzing in the blender.
3. Make a super-nutritious salad by throwing in some cooked grains to add texture, make it go further and fill you up.
4. Make your own muesli, granola or porridge, using a combination of processed whole grains.
5. Add some grains, such as pearl barley, to a stew or casserole – they will thicken and provide added nutrition.
6. Use a mixture of grains to accompany a meal rather than always cooking rice or potatoes.

Nutritional Composition of Grains

Grain (per 100g cooked)	Energy (kCal)	Protein (g)	% GDA	Fat (g)	% GDA	Carbohydrate (g)	% GDA	Fibre (g)	% GDA	Gluten-free?	Calcium (mg)	% RNI	Iron (mg)	% RNI	Magnesium (mg)	% RNI
Amaranth	102	3.8	6.9%	1.6	1.7%	18.7	6.2%	2.1	8.8%	Yes	47	6.7%	2.1	24.1%	65	21.7%
Pearl barley	123	2.3	4.2%	0.4	0.4%	28.2	9.4%	3.8	15.8%	No	11	1.6%	1.3	14.9%	22	7.3%
Buckwheat groats, roasted, cooked	92	3.4	6.2%	0.6	0.6%	19.9	6.6%	2.7	11.3%	Yes	7	1.0%	0.8	9.2%	51	17.0%
Bulgar wheat	83	3.1	5.6%	0.2	0.2%	18.6	6.2%	4.5	18.8%	No	10	1.4%	1	11.5%	32	10.7%
Cous cous	112	3.8	6.9%	0.2	0.2%	23.2	7.7%	1.4	5.8%	No	8	1.1%	0.4	4.6%	8	2.7%
Cous cous - wholewheat	173	6.0	10.9%	1.0	1.1%	32.0	10.7%	5.0	20.8%	No	Not available		Not available		Not available	
Kamut	146	6.5	11.8%	0.9	0.9%	30.5	10.2%	3.9	16.3%	No	10	1.4%	2	23.0%	56	18.7%
Millet	119	3.5	6.4%	1.0	1.1%	23.7	7.9%	1.3	5.4%	Yes	3	0.4%	0.6	6.9%	44	14.7%
Oats	71	2.5	4.5%	1.5	1.6%	12.0	4.0%	1.7	7.1%	No	9	1.3%	0.9	10.3%	27	9.0%
Polenta (cornmeal)	107	2.6	4.7%	0.6	0.6%	22.0	7.3%0	.9	3.8%	Yes	2	0.3%	0.2	2.3%	14	4.7%
Quinoa	120	4.4	8.0%	1.9	2.0%	21.3	7.1%	2.8	11.7%	Yes	17	2.4%	1.5	17.2%	64	21.3%
Rice - white, medium-grain	130	2.4	4.4%	0.2	0.2%	28.6	9.5%	0.3	1.3%	Yes	3	0.4%	1.5	17.2%	13	4.3%
Rice - white, long-grain	130	2.7	4.9%	0.3	0.3%	28.2	9.4%	0.4	1.7%	Yes	10	1.4%	1.2	13.8%	12	4.0%
Rice - brown, long-grain	111	2.6	4.7%	0.9	0.9%	23.0	7.7%	1.8	7.5%	Yes	10	1.4%	0.4	4.6%	43	14.3%
Rice - wild rice	101	4.0	7.3%	0.3	0.3%	21.3	7.1%	1.8	7.5%	Yes	3	0.4%	0.6	6.9%	32	10.7%
Spelt	127	5.5	10.0%	0.9	0.9%	26.4	8.8%	3.9	16.3%	No	10	1.4%	1.7	19.5%	49	16.3%
Teff	101	3.9	7.1%	0.7	0.7%	19.9	6.6%	2.8	11.7%	Yes	49.7	0%	2.1	24.1%	50	16.7%

Grain (per 100g uncooked, unless stated)

Grain	Energy (kCal)	Protein (g)	% GDA	Fat (g)	% GDA	Carbohydrate (g)	% GDA	Fibre (g)	% GDA	Gluten-free?	Calcium (mg)	% RNI	Iron (mg)	% RNI	Magnesium (mg)	% RNI
Buckwheat	343	13.3	24.2%	3.4	3.6%	71.5	23.8%	10.0	41.7%	Yes	18	2.6%	2.2	25.3%	231	77.0%
Cracked wheat	350	12.5	22.7%	1.3	1.4%	72.5	24.2%	12.5	52.1%	No	Not available		Not available		Not available	
Freekeh	351	12.6	22.9%	2.7	2.8%	72.0	24.0%	16.5	68.8%	No	53	7.6%	4.5	51.7%	110	36.7%
Oats	379	13.2	24.0%	6.5	6.8%	67.7	22.6%	10.1	42.1%	No	52	7.4%	4.3	49.4%	138	46.0%
Rye	338	10.3	18.7%	1.6	1.7%	75.9	25.3%	15.1	62.9%	No	24	3.4%	2.6	29.9%	110	36.7%
Chia seeds, dried	486	16.5	30.0%	30.7	32.3%	42.1	14.0%	34.4	143.3%	Yes	631	90.1%	7.7	88.5%	335	111.7%
Chia seeds, dried portion of 28.4g or 1oz	138	4.7	8.5%	8.7	9.2%	11.9	4.0%	9.8	40.8%	Yes	179	25.6%	2.2	25.3%	95	31.7%

Source: USDA National Nutrient Database for Standard Reference, Release 24 – 30/03/12
Polenta – Food Standards Australia and New Zealand NUTTAB 2010
Wholewheat couscous – Macro Wholefoods Market nutrition facts – www.myfitnesspal.com/food
Freekeh – Greenwheat Freekeh Nutritional Profile – www.greenwheat freekeh.com.au
Cracked Wheat – Bob's Red Mill Natural Foods – www.bobsredmill.com

Phosphorus (mg)	% RNI	Potassium (mg)	% RNI	Zinc (mg)	% RNI	Thiamin (mg)	% RNI	Riboflavin (mg)	% RNI	Niacin (mg)	% RNI	Vitamin B6 (mg)	% RNI	Folate (mcg)	% RNI	Vitamin E (mg)	Vitamin K (mcg)
148	26.9%	135	3.9%	0.9	9.1%	0.02	2.0%	0.02	1.5%	0.24	1.4%	0.11	7.9%	0	0.0%	0.2	0
54	9.8%	93	2.7%	0.8	8.4%	0.08	8.0%	0.06	4.6%	2.1	12.4%	0.1	7.1%	16	8.0%	0.01	0.8
70	12.7%	88	2.5%	0.6	6.3%	0.04	4.0%	0.04	3.1%	0.9	5.3%	0.1	7.1%	14	7.0%	0.09	1.9
40	7.3%	68	1.9%	0.6	6.3%	0.06	6.0%	0.03	2.3%	1	5.9%	0.08	5.7%	18	9.0%	0.01	0.5
22	4.0%	58	1.7%	0.3	3.2%	0.06	6.0%	0.03	2.3%	1	5.9%	0.05	3.6%	15	7.5%	0.1	0.1
Not available		Not available		Not available		Not available		Not available		Not available		Not available		Not available		Not available	
177	32.2%	202	5.8%	1.9	20.0%	0.1	10.0%	0.03	2.3%	2.7	15.9%	0.08	5.7%	12	6.0%	0	0
100	18.2%	62	1.8%	0.9	9.5%	0.1	10.0%	0.08	6.2%	1.3	7.6%	0.1	7.1%	19	9.5%	0.02	0.3
77	14.0%	70	2.0%	2.3	24.2%	0.01	1.0%	0.02	1.5%	0.2	1.2%	0.01	0.7%	6	3.0%	0.08	0.3
15	2.7%	49	1.4%	0.2	2.1%	0.2	20.0%	0.006	0.5%	0.1	0.6%	0.01	0.7%	11	5.5%	0.02	0
152	27.6%	172	4.9%	1.1	11.6%	0.1	10.0%	0.1	7.7%	0.4	2.4%	0.1	7.1%	42	21.0%	0.6	0
37	6.7%	29	0.8%	0.4	4.4%	0.2	20.0%	0.02	1.5%	1.8	10.6%	0.05	3.6%	97	48.5%	0	0
43	7.8%	35	1.0%	0.5	5.3%	0.2	20.0%	0.01	0.8%	1.5	8.8%	0.1	7.1%	97	48.5%	0.04	0
83	15.1%	43	1.2%	0.6	6.3%	0.1	10.0%	0.02	1.5%	1.5	8.8%	0.15	10.7%	4	2.0%	0.03	0.6
82	14.9%	101	2.9%	1.3	13.7%	0.05	5.0%	0.09	6.9%	1.3	7.6%	0.1	7.1%	26	13.0%	0.24	0.5
150	27.3%	143	4.1%	1.3	13.7%	0.1	10.0%	0.03	2.3%	2.6	15.3%	0.08	5.7%	13	6.5%	0.3	0
120	21.8%	107	3.1%	1.1	11.6%	0.2	20.0%	0.03	2.3%	0.9	5.3%	0.1	7.1%	18	9.0%	0	0
347	63.1%	460	13.1%	2.4	25.3%	0.1	10.0%	0.4	30.8%	7	41.2%	0.2	14.3%	30	15.0%	0	0
345	62.7%	405	11.6%	Not available		Not available		Not available		Not available		Not available		Not available		Not available	
Not available		440	12.6%	1.7	17.9%	0.4	40.0%	0.2	15.4%	Not available		Not available		Not available		0.4	0
410	74.5%	362	10.3%	3.6	37.9%	0.5	50.0%	0.2	15.4%	1.1	6.5%	0.1	7.1%	32	16.0%	0.4	2
332	60.4%	510	14.6%	2.7	27.9%	0.32	32.0%	0.25	19.2%	4.3	25.3%	0.3	21.4%	38	19.0%	0.9	5.9
860	156.4%	407	11.6%	4.6	48.4%	0.6	60.0%	0.2	15.4%	8.8	51.8%	0	0.0%	0	0.0%	0.5	0
244	44.4%	115	3.3%	1.3	13.7%	0.2	20.0%	0.05	3.8%	2.5	14.7%	0	0.0%	0	0.0%	0.1	0

Recommended Nutrient Intakes (RNIs) from Department of Health (1991) Dietary Reference Values for Food Energy and Nutrients for the UK. Report on Health and Social Subjects No.41. HMSO. London
There are no established UK RNIs for Vitamin E and Vitamin K
Guideline Daily Allowances (GDAs) from the Food and Drink Federation – www.gdalabel.org.uk
% GDAs are based on GDA values for men; GDAs for protein, carbohydrate and fat for women are slightly lower

Storing grains

Uncooked: Heat, light and air are the three things to consider when storing grains, as these will all affect their shelf life. You should store all uncooked grains in a cool, dark, dry place in a sealed container. Buy small quantities at a time – aim on having a two to three month turnover – and make sure new grains are put into clear containers; this will ensures freshness. Older grains will also take longer to cook. You can tell the freshness of grains by looking and smelling them – they should not have any sort of musty or rancid smell to them (it is the fat contained in the grains that turns rancid after time). Ensure flours and grains are kept in a sealed container. Not only is it better at preserving, but it will also prevent weevils (small beetles) from getting into the grains. If you do find weevils, throw away any contaminated food and start again.

Whole, 'unbroken' grains will keep for several months in a cupboard at room temperature or longer in the fridge or freezer. Refined and 'broken' grains, such as bulgar wheat or rolled oats, will spoil more quickly as they are essentially no longer sealed (which naturally preserves them for longer).

Cooked: Care should be taken when storing cooked grains, as some varieties, such as rice, can develop harmful bacteria, leading to food poisoning if they are not cooled and reheated carefully. Temperature is the most important factor, so it's best to serve rice and grains when they've just been cooked (though rice cookers are designed to keep rice at a safe temperature for an hour or so after cooking). If not using them straight away, cool the cooked grains within an hour and keep them refrigerated until needed; use within 24 hours. Throw away any rice and grains that have been left at room temperature overnight, and don't reheat rice and grains more than once.

Rinsing & toasting grains

Most grains should be rinsed thoroughly prior to cooking. Not only does rinsing remove any dirt, dust or agricultural sprays but it also, in the case of rice for example, removes starch from the grains to prevent them from becoming stodgy or 'starchy' during cooking. The easiest way to rinse grains is to put them in a pan or bowl, cover with plenty of water, then swirl with your hands before draining through a sieve, or carefully through your fingers. Repeat the process until the water runs clear, not cloudy; it can take two or three changes of water.

Soaking increases the moisture content of grains, speeds up cooking and, on the whole, give you a softer end result. Some grains require short soaking, for 10–20 minutes or so, others longer. Whole grains such as Kamut, wheat and rye berries, for example, should be soaked for as long as possible.

Toasting grains
To add an extra rich flavour, grains such as millet and bulgar wheat can be toasted before cooking. Simply place in a dry frying pan and gently heat, stirring, so that the grains evenly toast. Cook as usual.

My 'go to' quick-fix, five-step everyday salad

You can make eating well and healthily at lunchtime so much easier with a couple of packs of quick-cooking whole grains kept in the storecupboard. I then just raid the fridge in search of some colour and crunch (or whatever needs using up) and hey presto...

1. Cook some grains – wholegrain couscous, cracked or bulgar wheat, quinoa, buckwheat, wheat berries, spelt or pearl barley are my favourites.
2. Add some fruit/vegetables – cucumber, spring onion, radish, celery, peppers, lettuce, tomato, sweetcorn, asparagus, apple, avocado, grapes, grated carrot, pomegranate seeds, dried apricots, green beans, watercress, mangetout, rocket, pea shoots...
3. Throw in some crumbled feta, tinned tuna, cold crispy bacon, leftover roast or sausages, salami, prawns, goat's cheese, white beans or chickpeas.
4. Sprinkle over some nuts and seeds – pumpkin or chia seeds, toasted almonds, pine nuts, linseeds or cashews.
5. Drizzle with extra-virgin olive oil and lemon juice and season well.

Cooking Grains

Bear in mind that the cooking times for grains vary depending on the brand and the equipment used as well as the age of the grains. The following is therefore a guide and you should adjust the quantity of liquid and timings accordingly. When I say 'volume' measure I mean put it in a jug and see how many ml it comes up to and then use that as your volume.

Amaranth

Volume: 1 amaranth to approx. 3 water
Rinse thoroughly, then bring to the boil and simmer for 25 minutes or until tender, then drain. The amaranth will have a slightly porridge-like consistency after cooking.

Barley

Whole (Volume: 1 barley to approx. 4 water)
Rinse thoroughly then bring to the boil in plenty of water and simmer for 50–60 minutes or until soft.

Pearl (Volume: 1 barley to approx. 4 water)
Rinse thoroughly then bring to the boil in plenty of water and simmer for 20–30 minutes or until soft. Alternatively, add directly to soups and casseroles about 20 minutes before the end of cooking time.

Buckwheat

Volume: 1 buckwheat to approx. 2 water
When cooking raw, untoasted buckwheat, rinse thoroughly, then place in a pan and add the water. Bring to the boil then reduce the heat and simmer for 10 minutes or until the buckwheat is tender. Add a little extra water if required. Drain off any excess liquid before use.

Bulgar wheat

No-cook method: This method gives a slightly more nutty bite to the grains. Rinse 100g bulgar in three changes of water, swirling it around with your fingers, then drain and transfer to a bowl. Cover with 150ml boiling water or stock and cover with clingfilm. Leave for 1–1 ½ hours, or until it has reached the desired softness, then fluff up with a fork.

Stove-top method: This is my preferred method for most recipes as the end result is a little softer. Rinse 100g bulgar wheat in three changes of water, swirling it with your fingers, then drain, transfer to a pan and cover with 500ml boiling water or stock. Bring up to the boil, turn down the heat and simmer for 10–12 minutes or until tender but with a nutty bite, making sure that the grains have enough water so as not to catch. Rest, covered, for 10 minutes then fluff up with a fork.

Direct method: In addition, there is the simple method of adding the bulgar wheat directly to soups and stews, which should be done about 12 minutes before the end of cooking.

Chia seeds

There is no need to cook chia seeds, but grinding them before use or putting them in a blender (as per the smoothie recipe on page 51) is more nutritious.

You can also buy them cold-pressed: this not only has health benefits but also prevents unsightly chia seeds stuck in the teeth!

Couscous
Volume: 1 couscous to 1½ water or stock
No-cook method: Put the couscous into a bowl and pour over the boiling water or stock and seasoning, then cover and leave for 5–10 minutes or until the granules have swelled. (The time taken will vary if using quick-cook, barley or wholewheat couscous, so check the label). Add a dash of olive oil or a knob of butter, then fluff up with a fork.

Steaming method: This method is preferred for fluffier results, but is a little more time-consuming. Rinse the couscous in a sieve, then place in a bowl, just cover with cold water and leave to swell for 5 minutes. Mix with a dash of olive oil and some seasoning, then transfer into an oiled couscoussière or fine steamer and cook over boiling liquid for 20–25 minutes. Rest and fluff up with a fork before serving.

Direct method: If adding to soups or stews, simply sprinkle in the couscous 6–8 minutes before the end.

Giant couscous: Cook in boiling salted water for 8–10 minutes or until tender, then drain.

Cracked wheat
Volume: 1 cracked wheat to 2 water
Rinse the cracked wheat in three changes of water or until it runs clear. Drain thoroughly, put in a pan and cover with just boiled water. Bring to the boil, then gently boil for 10–12 minutes or until just tender but with a bite. Drain, then rest for 5 minutes. You can add the wheat directly to casseroles or soups.

Farro
Pearled and semi-pearled: when adding directly to soups, risottos or casseroles, pearled or semi-pearled farro takes 20–30 minutes to cook. If boiling, rinse the farro then cover with boiling water or stock and cook for 20–30 minutes.

Wholegrain: allow around 60 minutes, whether adding the farro directly to soups, etc. or cooking in boiling water or stock.

Freekeh
Volume: 1 freekeh to approx. 3 water or stock
Rinse the freekeh first then put into a pan with the boiling water (salted) or stock. Bring to the boil, cover and cook for 20–25 minutes or until the freekeh is cooked; make sure that there is enough water to cover the grains during cooking.

If using wholegrain freekeh, use a little more water and cook for 30–35 minutes.

Grits
Volume: 1 grits to 4 water or stock
Whisk the grits into the boiling liquid in a steady stream, then cover and cook for about 20 minutes, stirring once or twice, until soft. You may need to add a little extra water if the grits dry out too much.

Kamut   Volume: 1 Kamut to approx. 3 water
Rinse, then soak for 1 hour before cooking. Bring to the boil and cook for 1–1½ hours until tender but still chewy, adding more water if required.

Kasha   Volume: 1 kasha to approx. 2 water
Rinse throroughly, then add to soups and casseroles or bring to the boil in boiling water and simmer gently for 20–25 minutes or until tender, adding extra water if required.

Millet   Volume: 1 millet to approx. 3½ water
Bring to the boil and then simmer gently for 15–20 minutes or until tender; cook it for longer if you want a more porridge-like consistency. You can add pearled millet directly to muffins and cakes.

Oats   Rolled (Volume: 1 oats to approx. 2–2½ milk or water)
Cook for 10–15 minutes until creamy as a breakfast cereal, adding a little extra liquid if required, or add raw to baking recipes such as flapjacks or granola.

Quick-cook (Volume: 1 oats to approx. 2 milk or water)
Cook fine or quick-cook oats for 5–10 minutes or until creamy, adding extra liquid if required. They can also can be added directly to baking recipes.

Polenta   I use a mixture of stock and milk when making polenta. To serve four, pour 500ml chicken or vegetable stock and 200ml milk into a heavy-based pan and bring up to a simmer. Whisk in 100g cornmeal in a steady stream and continue to whisk for 2 minutes. Then put a lid on and cook over a very, very gentle heat for 1 hour, stirring every 10 minutes with a wooden spoon for a good 40 seconds or so to release the starch. If serving firm rather than creamy, pour the polenta into a container lined with clingfilm, cool completely and then cut into slices before grilling or frying and topping.

Quinoa   Volume: 1 quinoa to approx 2–2½ water or stock
Rinse the quinoa, then cook in stock or water for 9–12 minutes or until tender; black and red quinoa need to cook a little longer than the white. Drain off any excess liquid. If using in soups or stews, add 10 minutes before the end.

Rice   There are many methods for cooking rice out there. Some people wash the rice, then simply cover with plenty of boiling water and when the grains are soft, drain and fluff up. Some like to be more precise with their measurements, to ensure that the rice isn't drowned of flavour. If you cover the pan or if you don't, there's no wrong method, just what you've got used to and what you prefer. Personally, I prefer the covered method, which calls for a precise volume of liquid, as it gives a better texture to the rice and the end result is tastier and less watery, due to the minimal amount of liquid used. I have included the absorption method as it is a more traditional one, but it is a bit more of a faff. Easy-cook rice has been part cooked to make it faster and easier for you to prepare at home. I'm not a big fan, but in certain circumstances I'm sure it can be a quick fix!

Rice cookers are a great investment if you eat a lot of rice. All you need to do is put in the correct amount of rinsed rice and water (follow the instructions as the machines vary), and then leave the cooker to work its magic. The cooker will switch itself off when cooking is complete and will keep the rice warm for the next hour or so until you are ready to eat.

I sometimes add salt and sometimes don't – Basmati, brown and red rice often benefit from a little salt; Thai fragrant rice, on the other hand, doesn't need any salt as far as I'm concerned. And you don't have to cook rice in water; I think red rice, for example, is very tasty cooked in stock.

How much rice? A 400ml/400g measure of uncooked rice is enough for approximately four generous servings as an accompaniment to a curry or for a stir-fry; however, the quantity will of course depend on how much you like to serve (and eat!). The weight of rice approximately doubles once cooked, so if a recipe asks for 800g cooked and cooled rice, you will need to cook approximately 400g.

Covered method
White long-grain rice (includes Basmati and jasmine rice): rinse one measure of rice (eg. 1 cup) in two changes of water, swirling the rice with your hand. Drain, then re-wash. Boil the kettle. Put the drained rice into a pan and add 1½ measures (eg. 1½ cups water) of just-boiled water. Bring up to the boil, then cover with a tight-fitting lid and gently boil for 10 minutes. Turn the heat off, cover with a tea towel and lid and leave to stand for 5 minutes, before fluffing with a fork.

Brown long-grain rice: Rinse one measure of brown rice in two changes of water, swirling the rice with your hand, draining then re-washing. Boil the kettle. Put the drained rice into a pan and add 3½ measures of just-boiled water. Bring up to the boil, then cover with a tight-fitting lid and very gently boil for 35 minutes or until the rice is soft and the water all absorbed. Turn off the heat, cover with a tea towel and then the lid and leave to stand for 10 minutes, before fluffing with a fork and serving.

Absorption method
White long-grain rice (includes Basmati but not jasmine rice): rinse 400ml/400g long-grain rice twice in cold water, swirling it with your fingers. Drain, then cover with more cold water and leave to soak for 10 minutes. Drain again, then put the rice into a pan with 550ml cold water. Cover with a lid, bring to the boil, then, as soon as it's boiling, remove the lid and boil gently for 3–6 minutes until very nearly all the water has evaporated; the time taken will vary depending on the power of your cooker and the quantity of rice (larger quantities will take a little longer). Then, turn the heat to its lowest setting and continue to steam gently, covered with the lid again, for about 8 minutes. Turn off the heat, place a folded tea towel over the pan, followed by the lid, then leave to stand for 6–8 minutes before fluffing with a fork and serving.

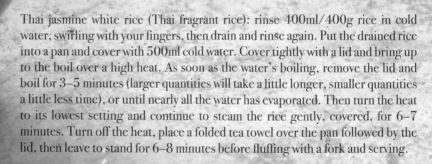

Thai jasmine white rice (Thai fragrant rice): rinse 400ml/400g rice in cold water, swirling with your fingers, then drain and rinse again. Put the drained rice into a pan and cover with 500ml cold water. Cover tightly with a lid and bring up to the boil over a high heat. As soon as the water's boiling, remove the lid and boil for 3–5 minutes (larger quantities will take a little longer, smaller quantities a little less time), or until nearly all the water has evaporated. Then turn the heat to its lowest setting and continue to steam the rice gently, covered, for 6–7 minutes. Turn off the heat, place a folded tea towel over the pan followed by the lid, then leave to stand for 6–8 minutes before fluffing with a fork and serving.

Brown long-grain rice: I like to soak brown rice as I think it gives the rice a better texture. However, if you are short on time, and want to cook it straight after rinsing, skip this stage and just add 125ml more water to the pan. Rinse 400ml/400g rice in cold water, swirling it with your fingers, then drain and rinse again. Cover with more water and soak for 1 hour. Put the drained rice into a pan and cover with 700ml cold water. Cover tightly with a lid and bring up to the boil, then immediately remove the lid and boil for 14–16 minutes or until nearly all the water has evaporated. Put the lid back on, lower the heat and steam gently for 8–9 minutes. Turn off the heat, place a folded tea towel over the pan, followed by the lid, then leave to stand for 6–8 minutes before fluffing up.

When using the absorption method and other non-covered methods of cooking rice, a crust can often form in the base of your rice pan during cooking. In Indonesia, this is called the *intip* and in Iran, the *tahdig*. In Spain, this crust is prized and is an essential part of a paella. If your rice develops a crust during the absorption method, leave the rice to rest with the lid on and this will loosen any grains. Or, follow tradition and enjoy its taste!

Ideal partners for long-grain rice
Flavours during cooking: use stock to cook the rice rather than water, or half coconut milk. Add fresh ginger, lemongrass, lime leaves, cinnamon or cardamom, lemon, lime or orange rind. Flavours after cooking: stir in chopped fresh herbs such as parsley, chives or basil. Add citrus zest and juice, toasted nuts or chopped dried fruit.

Cooking other types of rice
The quantities given here are sufficient for approximately two servings.

Wild rice
Wash 100g wild rice and transfer to a bowl. Cover with plenty of cold water and leave to soak for 1 hour (any longer will break the grains down too much). Drain, transfer to a pan and cover with 600ml cold water. Cover, bring slowly to the boil then remove the lid and boil gently over a medium heat until the rice is cooked – about 25–35 minutes, or until the grains have just opened but still retain a bite. Add extra boiling water if it's drying out. Drain, then return the rice to the pan and cover with a tea towel for 5 minutes. Fluff with a fork before serving.

Red rice

Wash 150g red rice and transfer to a bowl. Cover with plenty of cold water and leave to soak for 1–2 hours. Drain the rice, transfer to a pan with 550ml water. Cover with a lid, bring slowly to a boil, then remove the lid and boil gently over a medium heat until the rice is cooked – about 25–35 minutes, or until the grains are tender. Add extra boiling water in this time if the water is drying out. Drain off any excess water, return the rice to the pan and cover with a tea towel for 5 minutes. Fluff with a fork and serve.

Black rice

Wash 150g black rice and transfer to a bowl. Cover with plenty of cold water and leave to soak for 1–2 hours. Drain the rice, transfer to a pan with 600ml cold water. Cover with a lid, bring slowly to a boil, then remove the lid and boil gently over a medium heat until the rice is cooked – about 35–45 minutes or until the grains are tender. Add extra boiling water in this time if the water is drying out. Drain off any excess water, return the rice to the pan and cover with a tea towel for 5 minutes. Fluff with a fork and serve.

Other methods of cooking rice

Sometimes rice is cooked as part of a recipe, such as in the below recipes. Here are quick references to pages with descriptions of the most important classic rice dishes:

Paella: see page 179 – Pilaf: see page 105 - Rice pudding: see page 218 – Risotto: see page 94 – Sushi: see page 78

Rye Berries

Volume: 1 rye berries to approx. 4 water
Wash the rye berries thoroughly and cook in plenty of boiling water for 50–60 minutes, or until soft, topping up with water as required. Then drain.

Spelt

You can add rinsed pearled or whole spelt directly to soups, risottos or casseroles; allow 20–30 minutes cooking time for pearled or a further 20 minutes or so longer for whole. Alternatively, cover with boiling water or stock and cook until tender – 20–30 minutes for pearled, 40–60 minutes for whole.

Teff

Volume: 1 teff to approx. 3 water
Teff can be cooked as a cereal, rather like a porridge, or cooked and added to salads, dips and soups. Simmer in water for 15–20 minutes.

Wheat berries

Volume: 1 wheat berries to approx. 4 water
Rinse the berries thoroughly and cook in plenty of boiling water for 50–60 minutes, or until soft. Drain.

Breakfasts

Granola – DIY muesli – The ultimate hidden 'power' porridge – Spelt and millet muffins with apple and cinnamon – Soaked summer muesli – Salt and pepper oatcakes – Lontong (rice cakes) – Breakfast quinoa with raisins and honey – Ruby chia power juice – 'No knead' grain and seed loaf – Breakfast power smoothie – Lazy courgette & sun-dried tomato cornbread – Sweet and savoury cornmeal pancakes

Granola

'Granula' was invented in 1894 as a healthy breakfast by Dr. Connor Lacey at Jackson Sanitarium in Dansville, New York. (Around the same time a certain Dr Bircher-Benner was inventing his own healthy cereal in Switzerland!) What we now call granola has been reinvented a number of times since the original recipe, which was actually nothing like the sweet baked nuggets of grains, fruit and nuts that we now know and love. This recipe is the easiest and tastiest make-it-yourself breakfast granola, with no hidden nasties. I use a bag of four-grain porridge, available in some supermarkets, as the base – it's a mix of jumbo oats, flaked brown rice, flaked wheat and flaked barley. However, you can use just oats or a mix of the above. Also, feel free to up the nuts or change the fruit as per your tastes. Use the granola as a sprinkling for fruit compotes or eat with yogurt or milk. If you're keeping the granola for a couple of weeks, it's best to use oil rather than butter and properly dried fruits rather than the ready-to-eat dried fruits.

Preheat the oven to 160°C/140°C fan/gas mark 3.

In a bowl in the microwave or in a pan on the stove, melt the butter or oil with the honey and fruit juice. Add the remaining ingredients, except the dried fruit, and stir together until thoroughly combined. Pour onto a shallow baking tray in a layer approximately 2cm deep and place in the oven.

Bake for 20–30 minutes or until golden, turning the granola over a couple of times during the baking to ensure that it is evenly toasted (the outer edges tend to brown faster). Remove from the oven when the granola is golden brown and crisp. Stir in the dried fruit, then cool in the tray before transferring to an airtight container. Keep in a cool, dry place.

Serve with fresh fruit, yogurt or milk and a sprinkling of chia seeds, if you like.

Variations:

Apricot and pecan – add 100g chopped ready-to-eat dried apricots and 75g pecan nuts in place of the almonds and fruit.

Raisin and hazelnut – use 50g raisins and 75g hazelnuts in place of the dried fruit and nuts.

Maple and nut – swap the honey for maple syrup and replace the fruit with 75g halved pecans.

Serves 4–6

50g melted butter, or 2 tablespoons vegetable oil for a healthier version
3 tablespoons runny honey, preferably Manuka
3 tablespoons apple or orange juice
200g four-grain porridge mix, or rolled oats or a mix of grains
50g flaked almonds
1 heaped tablespoon pumpkin seeds
1 heaped tablespoon flaxseeds/linseeds,
50g dried apple
50g dried cranberries

DIY muesli

I am a total convert to home-made muesli and find it so much less sugary than the expensive shop-bought brands that often contain nothing but rather dusty-tasting grains and too many raisins. Plus, you can make it your own and vary it as you like – mixing your favourite grains and throwing in plenty of fruit and nuts. I like to lightly toast the grains as it gives them a less pappy taste and adds richness to the muesli – but if you prefer, just mix them from raw. Grain 'flakes' are cut a little thinner than just plain rolled, making them preferable for this recipe. You can buy a muesli base mix if you prefer – just use 350g of the base mix instead of the grains from the recipe – and simply toast it in the oven before adding the seeds, fruit, nuts, etc. Double the quantities if the whole family enjoy it – it will keep for a month or so in a cool dry place.

Preheat the oven to 180°C/160°C fan/gas mark 4.

Mix the grain flakes and bran (if using) together and spread out thinly (6–8mm thick) over a baking tray – you might need two. Bake for 6–8 minutes or until the grains feel lightly toasted (rather than browned) when you turn them over with a spoon. Once cool, transfer to a large bowl, add the remaining ingredients and stir together. Keep in an airtight container.

Variations:

To make the following variations, toast the grains as in the above recipe, then mix with the remaining ingredients and keep in an airtight container.

My favourite muesli – 100g barley flakes, 100g rye flakes, 150g oat flakes, 2 tablespoons wheat bran, 1½ tablespoons pumpkin seeds, 100g roughly chopped Brazil nuts, 100g mix of chopped dried apricots and raisins, a handful each of broken up dried banana chips and coconut shavings and 3 tablespoons chia seeds (optional).

Superfood muesli – 200g quinoa flakes, 200g oat flakes, 50g goji berries or dried blueberries, 50g chopped dates or prunes, 4 tablespoons chia seeds, 2 tablespoons flaxseeds/linseeds and 4 tablespoons toasted blanched almonds.

Gluten-free muesli – 100g uncontaminated gluten-free oats*, 200g quinoa flakes, 2 tablespoons flaxseeds/linseeds, 2 tablespoons sunflower seeds, 100g toasted and roughly chopped, whole unblanched almonds and 100g dried fruit.

* see page 235 for information about gluten-free oats

Serves 4–6

200g barley, quinoa, kamut, rye or spelt flakes, or a mixture
150g oat flakes or quick-cook oats
2 tablespoons wheat bran
3 tablespoons flaxseeds, linseeds, sunflower seeds, pumpkin seeds, sesame seeds or chia seeds
125g mixed dried fruits, e.g. chopped apricots or dates, sultanas, raisins, apple or coconut shavings, goji berries
100g mixed nuts, e.g. Brazil nuts, walnuts, cashews, almonds, pecans

Optional extras: chocolate chunks, dried banana pieces

The ultimate hidden 'power' porridge

Breakfast porridge, made from oats and simmered and served with sugar or salt, originated in Scotland. Traditionally, the oats were cooked and kept in a porridge drawer – then sliced and eaten by the crofters over a number of days. Cooking oats lowers their glycemic index, making you feel satisfied for longer than, say, a bowl of muesli. However, the addition of some extra grains and seeds makes it even more nutritious. You won't want to keep weighing out everything, so make a big batch of the grain mix and keep it in an airtight tin. About 50g of the mix will need approximately 300ml liquid for cooking – you can choose the ratio of milk to water according to how strict you want to be, but at least half and half is preferable for it to taste good. The quick-cook oatmeal adds an extra creaminess to the porridge. If you are not including the chia, teff and bran you may need less liquid.

Put all the grains and seeds in a pan and add the milk and 200ml cold water. Bring gently to a simmer, then cook for 10–15 minutes or until thick and creamy, adding the remaining 100ml water if the porridge is too thick. Pour into bowls and drizzle with honey. Serve sprinkled with fruit.

Serves 2–3

50g jumbo oats, triticale or a
 mix of both
25g quick-cook oats *
1 tablespoon flaxseeds/
 linseeds
1 tablespoon chia seeds
 (optional)
1 tablespoon wheat bran
1 tablespoon wholegrain teff
 (optional)
1 tablespoon pumpkin or
 sunflower seeds
350ml milk
300ml water
runny honey, preferably
 Manuka, to sweeten
fruit, e.g. fresh berries,
 banana or raisins, to serve

* see page 235 for a note about
uncontaminated, gluten-free oats

Spelt and millet muffins with apple and cinnamon

These are quite dense, full-flavoured muffins which are perfect for snacking on. Wholegrain spelt and millet are both packed with fibre, protein, vitamins and minerals and should be included more in our diets. Unlike flour made from common wheat, spelt contains soluble gluten, which can often be tolerated by those who are intolerant to ordinary wheat. Millet is mainly consumed in Africa and Asia, much less in the West. You can buy both wholegrain spelt flour and millet in health food shops. The two types of apple used in the muffins means that you get the texture of the chopped apple and intense apple flavour of the purée; you can make your own purée or just buy a jar.

Preheat the oven to 200°C/180°C fan/gas mark 6. Line a muffin tin with 12 cases.

Put the flour, millet, baking powder, salt, bicarbonate of soda, brown sugar and cinnamon into a large bowl and stir together. In another bowl or jug, mix the melted butter with the eggs, yogurt and apple purée. Add to the flour mixture and fold together until nearly combined. Add the chopped apple and walnuts or raisins and fold again gently until just combined.

Spoon into the muffin cases and top each with 2 apple slices. Bake for 20–25 minutes or until risen and springy to the touch. Leave to cool for 5 minutes before removing from the tin and dusting with the cinnamon icing sugar.

Makes 12 muffins

175g wholegrain spelt flour
25g hulled millet
1½ teaspoons baking powder
¼ teaspoon salt
1 teaspoon bicarbonate of soda
75g soft dark brown sugar
½ rounded teaspoon ground cinnamon
75g butter, melted
2 medium free-range eggs
3 tablespoons plain yogurt
150g sweetened apple purée or sauce
2 eating apples, peeled and cored, 1 chopped and 1 very thinly sliced
50g chopped walnuts or raisins
1 tablespoon icing sugar mixed with ½ level teaspoon ground cinnamon, for dusting

Millet

Latin name: Millet is the general name for many similar cereals of the genus Panicum

To many people in the West, millet means bird food. It may come as a surprise, therefore, that millet ranks as the sixth most important cereal grain in the world, sustaining more than a third of the world's population, particularly in developing countries. Millet is a hardy crop that grows with minimal intervention in less fertile soil and stores well; it can therefore be kept in reserve in case of famine or planted as a fall-back grain if other crops fail. India produces by far the highest quantity of millet, while many of the other leading producers are in African countries. High in fibre, protein, vitamins and minerals, millet could and should be included more in our own diets. Millet varies considerably in both look and taste. Some types are grown purely for animal fodder, while others are grown for human consumption. *Panicum miliaceum* or common millet (also known as hog, Indian or Proso millet) is the one most widely used in the West.

Millet was one of the first cereal grains to be domesticated, in the Neolithic era, and appears to have been cultivated simultaneously in Africa and Asia. It was the prevalent grain in China before rice, and the earliest written record of it, dating from 2800BC is Chinese; it gives detailed instructions for growing and storing the grain and lists it as one of the five sacred Chinese crops, along with soybeans, rice, wheat and barley. In fact, the first-known noodles were made from millet. The crop was a staple in arid areas of India and Africa for thousands of years and, once millet reached Europe (some time before 2000BC), it became the main staple there, grown even more widely than wheat.

While millet is no longer widely used by Europeans, it is still popular in parts of Eastern Europe. Millet porridge, for example, is traditional in Slovenia and Russia, just as it is in many countries in Africa and parts of Asia, including China. In western India, millet flour is mixed with sorghum to make the famous unleavened bread, roti, just as it has been for hundreds of years. Millet is also used to make millet beer in some cultures, including in Taiwan and Nepal. Awaokoshi, candied millet puffs, are a speciality of Osaka, Japan. Even though millet doesn't feature much in our cookbooks or menus, since the 1970s it has been gaining popularity in Western Europe and North America as a delicious, nutritious whole grain that is easy to cook. And there are numerous recipes from around the world to look to for inspiration as to how to use it.

Look: The colour of millet varies from cream to black. The most commonly eaten millet is pale in colour with tiny, bead-like seeds.
Taste and Texture: Boiled millet is mild and creamy tasting; you can toast the millet before cooking to create a nutty flavour.
Uses: Whole millet can be cooked and used in salads or added to stews. It is added to leavened bread and baking recipes for texture, or even popped like corn, as a snack. Ground millet can be cooked in water or milk (or both) to make a porridge-like cereal, best eaten sweetened, or used in unleavened breads.
Nutrients: Millet has a similar protein content to wheat and maize. It is gluten-free. It is high in soluble fibre and is relatively rich in iron and phosphorus, as well as B vitamins (especially niacin, B6 and folic acid), calcium, iron, potassium, magnesium and zinc.

Sorghum

Latin name: Sorghum bicolor

Other names: milo, guinea corn, jowari, durra, mtama and kafir corn

Sorghum is related and similar to millet, but is even more significant as a crop, being the world's fifth most important cereal. This is due largely to the fact that sorghum is naturally tolerant of drought and high temperatures and is also very versatile, being used as a foodstuff, animal feed and fuel. It is considered a staple food in the upland and drier parts of Africa and India.

There are many species and sub-species of sorghum, but *Sorghum bicolor* includes all cultivated varieties. The colour of the grains can vary from white and pale yellow to red, purple and brown; as a general rule, the white grains are used for food. Sorghum is different from other grains (including millet), in that it doesn't have an inedible outer hull, so can be eaten in its entirety. It is also gluten-free. *Saccharatum* cultivars (sweet sorghum) are not suitable for eating as grain, but the sap produced in their thick stems is used as a sweet syrup, with a flavour like molasses. It is popular in the southern United States, used on pancakes, grits and so on.

Sorghum is indigenous to Africa and was first cultivated in Ethiopia around 4000–3000BC, moving from there across to West Africa, the Middle East, India and China. Sorghum was introduced to the United States from Africa in the early 17th century. The US, Asia and nations of West Africa are now all major producers of sorghum.

While sorghum is used principally for animal feed and fuel in the West, the grains and flour are used widely in the kitchens of the less developed regions of Africa and Asia to produce couscous, bread and porridge. In South Africa, a popular chocolate-like porridge called 'maltabella' is made from sorghum, and in India, it is cooked like rice, in a dish called 'kichuri'. As it is gluten-free, sorghum flour is becoming increasingly popular in both commercial and home-made breads and is often one of the grains used in gluten-free flour. It can also be popped like maize.

Look: Small, pale, bead-like grains.
Taste and Texture: Neutral, slightly sweet in flavour, chewy in texture.
Uses: Sorghum grains can be cooked and used in side dishes and salads, while the flour can be used in pancakes, flatbreads and also porridge. Use it either as the only flour or added to regular wheat flour in order to add nutrition and fibre; the flour has a neutral taste so it does not affect the flavour of your foods like some other gluten-free flours.
Nutrients: Being gluten-free is one of sorghum's greatest nutritional plus points. It also contains calcium, potassium, magnesium and protein. Some varieties of sorghum are also high in antioxidants. The wax surrounding the sorghum grain contains compounds called policosanols, which may be beneficial for cardiac health.

Soaked summer muesli

This is a wonderful hand-me-down recipe for a bircher-style creamy muesli – oats soaked in fruit juice, to soften, then mixed with yogurt, fruit and whatever else you fancy. This recipe is from my friend Becca. It's ideal for warmer mornings when you feel like a fresh and lively start to your day. The recipe for muesli was devised by Dr Bircher-Benner for his Swiss patients in the 1890s. An advocate of healthy eating along with plenty of sleep and exercise, Dr Bircher was ahead of his time, prescribing a diet of raw fruit and vegetables and less white bread and meat for his patients. This version uses a mixture of fruit juice and milk to soften the oats, but it's a recipe easily adapted to suit your taste, so you can add only fruit juice if you prefer. You can also soak the oats for less time – an hour or so – if you want a speedier version.

Put the oats or porridge mix into a bowl and stir in the apple juice and milk. Leave to soak overnight in the fridge.

In the morning, stir the nuts, dried fruit and chia seeds (if using) into the oats, along with the yogurt. Add the fresh fruits of your choice, drizzle with honey if you like, and serve.

Serves 3

100g porridge oats or a porridge mix of oats, wheat and barley flakes and brown rice flakes (found in some supermarkets)
150ml apple juice
100ml milk
2 tablespoons mixed chopped nuts and/or dried fruit such as sultanas or chopped apricots
2 tablespoons chia seeds (optional)
4 heaped tablespoons creamy Greek, plain or fruit yogurt

To serve:
fresh fruit of your choice, e.g. 1 apple, peeled and grated, 1 sliced banana, a handful of summer berries, 1 peach or nectarine, sliced
honey (optional)

Salt and pepper oatcakes

Scottish oatcakes, traditionally made either on a griddle or baked, were, for centuries, eaten with every meal. It was even common practice, as far back as the 14th century, for the chieftain and the rest of his clan to travel with a small sack of oatmeal when heading into battle; they mixed the oatmeal with water and baked the biscuits over the campfire using their shields as iron plates. From the 19th century onwards, the savoury biscuits were commonly served to accompany soups, meat and fish dishes. This recipe is packed full of oats and produces biscuits with a long shelf life – which means that you can bake a batch and not feel you have to eat them in one sitting! You can omit the cheese or add some very finely chopped rosemary, chilli powder, sun-dried tomatoes or black olives, if you like. You can also use wholewheat flour instead of plain (in which case omit the wheat bran).

Preheat the oven to 180°C/160°C fan/gas mark 4.

Put all the dry ingredients, cheese and a good grind of pepper into a bowl and stir together. Make a well in the centre and add the olive oil, combining it with the dry mixture using a knife, then rub together using your fingertips. Add 5–6 tablespoonfuls of water and, again using the knife, mix together, then form into a ball using your hands.

Roll out on a floured surface to about 5mm thick, then stamp out circles of whatever size you like – I prefer them a little smaller than the shop-bought size, so use a 6cm cutter. Bake for 20–25 minutes or until the oatcakes are completely crisp. When cool, transfer to an airtight container.

Gluten-free alternative: you can make these oatcakes gluten-free by using uncontaminated oat flour (see page 235) and oat bran instead of plain flour and wheat bran. Follow the recipe as above.

Makes about
25 small biscuits

100g plain flour
125g quick-cook
   porridge oats
2 tablespoons wheat bran
½ teaspoon sea salt flakes
scant ½ teaspoon sugar
¾ teaspoon baking powder
2 teaspoons flaxseeds or
   linseeds, preferably
   partially ground
2 teaspoons chia or
   poppy seeds
2 teaspoons sesame seeds
1½ tablespoons grated
   Parmesan (optional)
3 tablespoons olive oil
freshly ground black pepper

Lontong (rice cakes)

In parts of Asia, rice is steamed in banana or coconut leaves to form a firm but moist cylindrical cake of compressed rice called 'lontong'; it is particularly common in Indonesia and Malaysia. The cake is cut into slices and eaten on its own or with satay, or is added to soups or to salads such as *gado gado* (see page 137). A version of lontong are also served for breakfast in the Phillipines, where it is known as *suman*.

Put the rice into a bowl, cover with water, swirl with your hand, then drain. Repeat the process then cover the rice with more cold water to 2cm above the top of the rice. Stir in the salt and leave overnight to soak. The next day, drain off nearly all of the water (leaving about 1 tablespoonful behind).

Carefully, hold the banana leaves over a flame (shiny-side down) and heat them, moving the leaves over the flame until they turn brighter in colour – this softens them, making them easier to roll. Be careful not to burn the leaves or your fingers! Roll up the banana leaves, one at a time, so they form cylinders 4–5cm in diameter. Using a torn-off strip of leaf or some cotton, tie a knot around the base, to seal one end.

Spoon in the rice until the cylinder is three-quarters full, squashing the rice down slightly. Seal the end, leaving a gap of about 3cm at the top to allow the rice to expand as it cooks. Repeat with the remaining banana leaves.

Place the rice parcels upright in a pan and cover completely with cold water. Cover with a lid and bring to the boil. Turn the heat down to a simmer and cook for 40 minutes. Drain off the water and leave the parcels to cool completely. Chill in the fridge until ready to use, then unwrap and slice.

Serves 4–6

400g Thai fragrant rice
¼ teaspoon salt
4 pieces of fresh banana leaf,
    each measuring
    approximately 22 x 28cm

Nasi kuning

At street stalls in Indonesia in the mornings you will often find *nasi kuning* being sold in cones for breakfast. This rice flavoured with turmeric and often coconut milk, along with *nasi champor* (plain rice with meat and vegetables), is eaten every day. For birthdays and celebrations, *nasi kuning* is often served with soy beans, meat, eggs and chicken and a fiery sambal alongside.

Breakfast quinoa with raisins and honey

Personally, I feel that quinoa when cooked by itself as a porridge is not that exciting. However, I am a big fan of the expression 'whatever floats your boat', and in the case of this recipe if you like the fresher taste of pure quinoa, then go for it and don't add the oats as I've suggested. For me, this combination is tasty as well as good for you, and my children adore it for breakfast. I have used cups in this recipe – when measuring early in the morning, why not keep it simple?

Put the quinoa, oats, chia (if using) and milk or milk/water combo into a pan and bring up to a simmer. Cook over a gentle heat for 8–10 minutes, stirring every so often, until just thickened. Add the raisins for the final 2–3 minutes; if using fresh fruit, just sprinkle it onto the bowls of porridge once cooked. Serve straight away, drizzled with honey.

Serves 2 generously

½ cup or small mug quinoa flakes, rinsed in a seive
½ cup or small mug quick-cook oats *
2 tablespoons chia seeds, preferably ground (optional)
3 cups milk (or 2 cups milk and 1 cup water)
a handful of raisins or summer fruits
runny honey, preferably Manuka

* see page 235 for a note about uncontaminated, gluten-free oats

---

Ruby chia power juice

Originally grown by the Aztecs thousands of years ago, chia seeds are tiny nuggets of pure power – packed with omega-3 fatty acids, antioxidants, fibre, protein and phytonutrients. The size of a poppy seed, they swell up to nine times their size when added to liquid and so keep you full for longer. The easiest way to include them is to either sprinkle a spoonful on your cereal, add them to porridge or make a power juice such as this one. I prefer it to the better known 'chia fresca', which consists simply of chia seeds, water and fresh lemon juice. Feel free to adjust the fruit to what you have and do buy frozen bags of fruit so you can have a stock always available.

Put all the ingredients into a blender. Whizz for a minute to thoroughly chop and blend, then serve.

Makes 300ml (enough for 1)

50g raspberries
seeds from ½ pomegranate or 100ml fresh pomegranate juice
4 large ripe strawberries
juice of 1 large orange
1½ tablespoons chia seeds
1 tablespoon wheat or oat bran
a little honey (optional)
3 or 4 ice cubes

'No knead' grain and seed loaf

Bar a few good Aussie bakeries dotted around, the bread in Singapore (where I live) is not massively interesting or healthy, so it's great to have a recipe for a cheat's loaf up your sleeve. This one is my Swedish friend Anna's creation, but adapted slightly to suit my family's likes. It's quite a dense bread, but certainly less dense and softer than darker pumpernickel-style breads, and uses wholegrain spelt and wholemeal flour rather than rye flour, which is commonly used in Scandinavian and German breads. It's the perfect bread to quickly rustle up too, as it takes half the time a normal loaf takes to make, and it's a great accompaniment to a bowl of soup, a smear of thick-set honey or some smoked salmon, chopped dill and lemon.

First, grease two large 900g loaf tins with oil.

Mix the flour, spelt flour, seeds, sugar, salt and yeast in a large bowl. Make a well in the centre and pour in 850ml warm water and the oil. Mix well with a spoon or your hand (it will be quite a wet mixture compared to normal bread dough).

Divide between the two greased loaf tins. Cover with a damp tea towel or an upturned mixing bowl and then leave in a warm place (the airing cupboard is perfect) for 1–1½ hours until it has nearly doubled in size.

Preheat the oven to 180°C/160°C fan/gas mark 4.

Uncover the loaves and bake for 40 minutes until golden and risen. Tip the bread out of the tins and return to the oven rack for another 10–15 minutes – then test by tapping the base to hear if it sounds hollow. If it doesn't, return to the oven for a further 5–10 minutes or until done. Leave to cool completely on a wire rack before slicing.

Alternative to spelt flour: instant oatmeal or use all stoneground wholewheat flour

Makes 2 x 900g loaves

3 tablespoons light olive or vegetable oil, plus a little for greasing

700g stoneground wholewheat flour

200g wholegrain spelt flour

4 tablespoons pumpkin seeds

3 tablespoons flaxseeds/ linseeds, preferably ground

3 teaspoons caster sugar

3 teaspoons sea salt flakes

7g sachet fast-action yeast

Breakfast power smoothie

The Aussies are big smoothie fans and back in the '90s I remember trying my first banana and strawberry smoothie in a Sydney café – a world of difference to the unhealthy banana syrup milkshakes or ice cream sodas that we would occasionally make as a childhood treat!

Put all the ingredients into a blender and whizz for a minute or until smooth. Pour into one or two glasses.

Alternative to wheat bran: oat bran; use uncontaminated oats or oat bran for a gluten-free version *.

* see page 235 for a note about uncontaminated, gluten-free oats.

**Makes 400ml
(enough for 1–2)**

150ml low-fat milk
3 heaped tablespoons yogurt
1 banana, roughly broken up
5 large ripe strawberries
½ tablespoon wheat bran
1 heaped tablespoon fine
    porridge oats
3 ice cubes
1 tablespoon chia seeds
    (optional)
a little honey (optional)

Lazy courgette and sun-dried tomato cornbread

Native Americans were the first to grow and grind maize, and they made a basic cornbread using nothing but ground corn, salt and water. Because of the different varieties of corn grown throughout North America, cornbread differed by region: blue corn was popular in the southwest, while the northern regions favoured yellow corn, and the south white corn. Over time, cornbread recipes also developed, including sweetened versions using molasses or honey. Nowadays, cornbread is still extremely popular in the United States and versions can be found all around the world. It can be served on its own or as an accompaniment to dishes. This recipe is a sort of traybake bread that begs for a bowl of home-made soup to be dipped into. If you can't find fine cornmeal, then whizz medium-ground cornmeal in a blender to break it up a little more. Alternatively, leave the medium cornmeal as it is for a grainy texture. You may find it easier to source polenta than cornmeal: if so, use polenta, as they are one and the same.

First, in a bowl, mix the buttermilk and cornmeal and leave to stand for 30 minutes in a cool place. Preheat the oven to 170°C/150°C fan/gas mark 3 and grease a square 18 x 18cm baking tin.

Heat the oil in a pan and fry the onion gently until soft – about 10 minutes. In a large bowl, sieve in the flour, then add the sea salt flakes, baking powder, bicarbonate of soda and sugar and mix together lightly.

Add the eggs to the buttermilk mixture and stir in the melted butter, onion, sun-dried tomatoes, grated courgette, sweetcorn and chilli flakes as well as a grinding of pepper. Add to the flour mixture and fold together using a large metal spoon until just combined. Pour into the greased tin. Bake for 45 minutes or until a skewer comes out clean. Then cool slightly, before turning out onto a board and cutting into squares.

Makes 9 squares

275ml buttermilk
175g fine cornmeal
  or polenta
1 tablespoon flavourless oil
1 medium onion, chopped
275g self-raising flour
¾ teaspoon sea salt flakes
1 teaspoon baking powder
1 teaspoon bicarbonate
  of soda
1½ teaspoons caster sugar
2 medium free-range eggs
75g butter, melted
8 sun-dried tomatoes,
  chopped
½ large courgette, grated
5 heaped tablespoons
  cooked sweetcorn kernels,
  either fresh or tinned
½ teaspoon chilli flakes
freshly ground black pepper

Sweet and savoury cornmeal pancakes

These simple cornmeal breakfast pancakes are known in the US as 'johnnycakes'. Apparently the settlers of New England were taught by the local Pawtuxet Indians how to grind and use corn for eating. In our house they've been re-christened 'Wilbo cakes' as my son's favourite job at the weekend is to whisk up the pancake batter. They are healthier and far tastier than the usual American pancake as they contain whole corn, ground, but if you don't like the grittiness then you can use a finer-ground corn or just use 50g extra flour and omit the corn. Many recipes use buttermilk, but as I wanted these to be an easy 'turn to' recipe, I have just added vinegar to the milk, which will do the same job. Note that there are small variations in the ingredients for the sweet and savoury versions. If making sweet pancakes, top with some melted butter and maple syrup. For the savoury version, add some crispy bacon.

In a jug, stir the vinegar into the milk and leave for 5 minutes on the worktop. Then add the cornmeal and leave to rest in the fridge for a further 25 minutes. Sieve the flour and bicarbonate of soda into a large bowl and add the salt and the sugar, if using. Whisk the eggs and melted butter thoroughly into the cornmeal mixture, then make a well in the flour and pour in the cornmeal mixture. Whisk together. You must then use the mixture straight away.

Melt some more butter in a non-stick frying pan and add 2 dessertspoonfuls of the batter per pancake. Fry over a low-medium heat until small air bubbles appear on the surface of the pancakes and then flip over. Cook until both sides are golden, then keep warm in a low oven while you cook the rest.

Variations:
Sweet blueberry pancakes with maple-orange butter – Mix 75g slightly softened butter with a little maple syrup and the zest of ½ an orange then chill in the fridge. Add a handful of fresh/frozen blueberries to the sweet pancake mixture and fry as above. Top with the maple-orange butter.

Sweetcorn pancakes with crispy bacon and roasted tomatoes – Add 4 table-spoonfuls of drained corn kernels and a sprinkling of cayenne pepper to the savoury mixture. Top the fried pancakes with crisp streaky bacon rashers and some roasted tomatoes.

Makes 12–15 pancakes

1½ teaspoons mild vinegar
275ml milk
50g fine or medium cornmeal or polenta
225g self-raising flour
1 teaspoon bicarbonate of soda
a pinch of sea salt flakes for sweet or ½ teaspoon for savoury pancakes
3 tablespoons sugar (sweet pancakes only)
2 free-range eggs, beaten
2 tablespoons melted butter, plus extra for frying

Soups & starters

Tuscan vegetable and farro soup – Middle Eastern pumpkin soup with kasha – Goan fish and rice soup – Wholesome chicken, vegetable, corn and barley broth – Spicy avocado and quinoa salsa – Babaganoush with teff – Griddled polenta with spicy sweet and sour aubergine – Spiced sweetcorn and pumpkin fritters – Mini arancini – Mini smoked salmon coulibiac parcels – Sushi 'my way' (maki and California rolls – Nigiri with tuna or salmon)

Tuscan vegetable and farro soup

Originating from the Tuscan town of Lucca, this soup is a wonderful sight on the table – its colours and smells are so rich and inviting! It's the ideal soup to serve for a weekend lunch as it's filling and needs only some crusty bread. The soup uses Italian farro, which is tricky to define as it can embrace a variety of ancient grains, including emmer wheat and spelt. If you want to use wholegrain farro for this you can, but you will need to cook it for nearer to one hour rather than half an hour, so some extra stock will be needed. You can also substitute dried soaked borlotti beans for the tinned version used here, especially if you're doing a slower-cooked version as they will take longer to cook. If you haven't got fresh stock, I like to use slightly watered-down tinned beef consommé for this sort of thing, as I think it has way more flavour than cubed stock. If you have some fresh pesto to hand, that is also a lovely addition – just top each bowl with a spoonful before serving.

In a large saucepan, gently fry the onion, celery and carrot in the oil for about 5 minutes. Add the pancetta or bacon and cook for a further 5 minutes. Stir in the garlic and farro grains, stir for a minute or so, then add the chopped tomatoes, boiling hot stock, thyme, bay and sugar. Season, bring to the boil, then cover and simmer for about 30 minutes. Add the drained and rinsed beans and cook for a further 10 minutes.

Taste for seasoning, adding extra if needed, then serve each bowl with a drizzle of extra-virgin olive oil, fresh basil and grated Parmesan sprinkled over the top.

Alternative to farro: pearl barley or pearled spelt

Serves 4–6

1 onion, chopped
2 sticks of celery, finely chopped
2 carrots, finely chopped
2 tablespoons extra-virgin olive oil
100g cubed pancetta, or chopped streaky bacon
2 garlic cloves, chopped
75g semi-pearled or pearled farro
1 x 400g tin chopped tomatoes
700ml beef stock
2 sprigs of thyme
1 bay leaf
a pinch or two of sugar
1 x 400g tin borlotti beans, drained

To serve:
olive oil, for drizzling
plenty of chopped basil
freshly grated Parmesan

Middle Eastern pumpkin soup with kasha

The word *kasha* in Eastern European cooking refers to a meal prepared using any grain, but generally in English 'kasha' refers to toasted buckwheat groats, which is what I have used in this recipe. While not necessarily a grain commonly associated with Middle Eastern cooking, kasha is a great choice to stand up to the spices and hearty flavours of the pumpkin. The combination of superfood powerhouse amaranth with kasha (or bulgar wheat), lots of vegetables and Middle Eastern spices makes this a seriously healthy and nutritionally balanced soup. It really hits the spot on those days when you need a bit of a warm and spicy lift!

Preheat the oven to 180°C/160°C fan/gas mark 4.

Scatter the pumpkin pieces and chilli over a baking tray, pour over half the oil and season with salt and pepper. Toss together, then bake in the oven for 30–35 minutes or until the pumpkin is tender. Remove from the oven and set aside.

Meanwhile, heat the remaining oil and butter in a saucepan and gently fry the onion and leeks for 10 minutes or so until soft. Stir in the garlic and ginger and cook for 2–3 minutes before adding the cumin, cinnamon, sumac (if using), amaranth and kasha. Stir over the heat for a further 1 or 2 minutes. Add the boiling hot stock and tomatoes and bring up to the boil, then simmer for 20–25 minutes, stirring occasionally.

Chop the roasted chilli and stir into the soup along with the pumpkin and a squeeze of lemon. Serve garnished with the coriander, a drizzle of yogurt and some sumac (optional).

———————

Alternative to kasha: buckwheat groats or bulgar wheat (the latter must be washed thoroughly in cold water, drained and washed until the water runs clear)

Serves 4–6

800g pumpkin (approximately 650g once peeled and deseeded), cut into 2cm cubes
1 large red chilli, halved lengthways and deseeded (unless you like it more fiery)
2 tablespoons olive oil
a knob of butter
1 onion, chopped
1 leek, chopped
3 garlic cloves, chopped
4cm piece fresh ginger, chopped
½ teaspoon cumin
2 pinches of cinnamon
½ teaspoon sumac (optional)
50g amaranth grains
50g kasha, washed
1 litre chicken or vegetable stock
1 x 400g tin chopped tomatoes
juice of ½ lemon
sea salt and freshly ground black pepper

To serve:
a generous handful of coriander, chopped
2–3 tablespoons natural yogurt seasoned with a little salt and pepper

Goan fish and rice soup

This is a soup-cum-stew-cum-curry! Certainly a main course in a bowl as opposed to a delicate starter, it can be eaten as a hearty lunch or a comforting supper. Featuring seafood, coconut milk, rice and local spices, all foods associated with Goan cuisine, it's rather like an Indian version of a chowder. I love the fact that the rice has the chance to soak up all the flavours, and is then piled into one comforting big bowl. There are many variations out there, mostly because of the type of fish chosen and subtle changes with the spices used. You can omit the prawns or scallops and use just fish if you prefer, in which case add another 150g or so. If you want a milder chilli flavour, deseed the chillies before adding.

Soak the rice in cold water for 10 minutes, then drain.

Meanwhile, fry the onion in a deep frying pan over a low heat for 6–8 minutes in the oil until golden. Add the garlic, ginger, garam masala, mustard seeds, curry leaves and chillies and stir well for a minute. Then add the tomatoes, vinegar, sugar, boiling hot stock or water and salt. Bring up to a simmer and cook for 2–3 minutes.

Add the coconut cream, bring up to the boil and stir in the rice and turmeric. Simmer for 5 minutes, then stir in the fish and seafood and continue to cook for a further 5 minutes or until the fish is cooked through and the rice is tender. Remove the chillies before serving.

Squeeze in the lime juice and add half the coriander. Serve in bowls with the remaining coriander sprinkled on top and lime wedges on the side.

_____

Serves 4

150g basmati rice
1 medium onion, sliced
1½ tablespoons vegetable or nut oil
2 garlic cloves, chopped
5cm piece fresh ginger, chopped
1½ teaspoons good-quality garam masala
½ teaspoon mustard seeds
20 small curry leaves
2 whole small red chillies, split
2 tomatoes, chopped
2 teaspoons white wine vinegar
2 teaspoons palm or brown sugar
650ml fish stock or water
½ teaspoon salt
400ml coconut cream
1 teaspoon turmeric
500g white fish, cut into chunks
12 prawns or 6 large scallops, halved
juice of 1 lime, plus wedges to garnish
a good handful of coriander, chopped

Cypriot trahana soup

Trahana is made from a mixture of coarse bulgar wheat known in Cyprus as *pourgouri* and sour milk or yogurt, and can be bought as dried 'chips' or pieces from any good Cypriot grocer. Traditionally, it was made in the villages of Cyprus whenever there was milk to spare – the fermented milk and *pourgouri* mixture was, at the last stage of making, dried in the sun and then tied up in a cloth and hung from a hook in the ceiling to keep until needed. Nowadays, it is machine made, but people will still seek out hand-made or village-made trahana (the quality depending on the maker), with which to make a good and warming winter soup with stock and halloumi cheese.

Wholesome chicken, vegetable, corn and barley broth

I've been in a quandary. Should you use a whole chicken to make a chicken soup or just use stock from a leftover carcass? I deliberated for a while and then remembered my Scottish grandmother, with her incredible knack of being able to turn very little into a feast. So, I've decided that a soup like this should be a thrifty triumph. However, if you want to simmer a whole chicken for a couple of hours with some veg and water to make a stock from scratch, or throw in a chicken leg to poach in the broth below and then strip flesh from the bones, or simply add some leftover roast chicken to the soup, then you can do any of the above with my blessing – but maybe not my grandmother's! Pearl barley has had its outer bran layer polished off so is lower in fibre than wholegrain barley, but the refined grains have a wonderful silky yet slightly chewy texture.

Heat the oil in a large pan, add the celery and carrot and gently soften for 8–10 minutes. Add the leeks, tomatoes, garlic and barley and stir-fry for 2 minutes before adding the boiling hot stock and soy sauce. Season generously. Bring gently up to the boil and simmer for 30 minutes, covered, or until the carrot and barley are nearly tender. Add the corn and simmer for a further 10 minutes.

Taste for seasoning, stir in the parsley and serve with hot garlic bread.

_____

Alternative to pearl barley: pearled spelt

Serves 4

2 teaspoons olive oil
2 sticks celery, trimmed and finely chopped
2 carrots, chopped
2 small leeks (or 1 large), trimmed and chopped
2 small tomatoes, cut into quarters
2 large garlic cloves, chopped
75g pearl barley
1.2 litres well-flavoured home-made chicken stock
2 teaspoons dark soy sauce
200g corn kernels (2 small cobs), uncooked, or 200g drained corn from a tin
1 tablespoon chopped flat-leaf parsley
sea salt and freshly ground black pepper

To serve:
garlic bread

Barley
Latin name:
Hordeum vulgare

A member of the grass family, barley evolved from its wild relative *Hordeum vulgare* subsp. *spontaneum*, which still grows wild in the Middle East. Barley was one of the founding agricultural crops of the Fertile Crescent, and is thought to have been first domesticated in Mesopotamia (modern-day Iraq) in around 8500BC. It is known that barley was one of the staple cereals of the ancient Egyptians, who buried mummies with necklaces made with grains. Barley was one of the major crops that accompanied the spread of agriculture into Europe during the 6th and 5th millennia BC. It became, along with rye, one of the peasantry's staple grains, used in breads, soups and stews – though in 1324 Edward II of England wasn't against using it to standardise the inch as equal to 'three grains of barley, dry and round, placed end to end lengthwise.'

Barley can be cultivated in temperate climates. It does not need especially rich soils so can be grown almost everywhere. Russia produces the highest quantity in the world, followed by the Ukraine and Canada. It is used all over the world, as animal feed and a source of malt as well as a cooking ingredient. In Japan and Korea, where barley has been cultivated for thousands of years, the grain is an ingredient in several traditional foods and drinks, including tea and miso, and is also popular mixed with rice.

In its whole form, barley is available as 'hulled' (with just the husk removed) or 'pearled' (with the husk and some or all of the outer bran layer polished off). Hulled or wholegrain barley is the most nutritious, as the bran layer remains intact, but pearled barley is more popular since it cooks in half the time and has a better texture. Pearl barley can be polished to varying degrees, but note that the more it's refined the less nutritious it is, though this is of course relative. Semi-pearled or Scotch barley has undergone less polishing so the bran layer remains partly intact, producing a nuttier texture. Look out, too, for intriguing black barley (also known as purple hull-less barley), which can be bought online. It derives from an Ethiopian variety and is now grown in the US. It is the only grain that can go from field to table without being processed as the bran layer stays attached to the kernel and is edible. It has a longer cooking time than regular barley (1–1½ hours), and can be used in side dishes and salads.

You can also buy barley grits, toasted and cracked barley grains, barley flakes and barley flour. Paula Wolfert, in her book *Greens and Grains*, describes making her own barley couscous by steaming the grits, though barley couscous can now be bought off the shelf too.

Look: Whole barley grains are oval and pale brown in colour, similar to wheat berries. A lighter shade if semi-pearled, creamy coloured if fully pearled. Barley flakes resemble porridge oats but are larger and darker in colour.
Taste and Texture: Barley tastes mild, nutty and rich, almost honey-like. Pearl barley has a particularly wonderful silky texture.
Uses: Throwing some barley into soups, stews and casseroles is a fabulous way to add some nutritious heartiness. Try using whole or pearled barley in salads, risottos and pilafs. Use barley flakes in muesli.

Nutrients: Barley is an excellent source of soluble fibre, which can help to lower blood cholesterol levels. Hulled barley contains high levels of calcium, magnesium, phosphorus, potassium, vitamin A, vitamin E, niacin and folate and is high in thiamin. Pearl barley contains smaller quantities of most of these nutrients and of course less fibre, yet remains fairly nutritious.

Teff
Latin name:
Eragrostis tef

Other names: tef, ttheff, tteff,
thaff, tcheff, thaft and taff!
And Ethiopian millet

The smallest grain in the world, tiny teff is the only fully domesticated member of the genus *Eragrostis* (lovegrass). It is thought to have originated in Ethiopia between 4000 and 1000BC.

Teff varieties vary in colour from white and red to dark brown. A small amount of the seed produces a mass of plants – one pound of teff sown can produce up to one tonne of the tiny grains in only 12 weeks. It is largely pest- and disease-resistant and thrives in harsh and unpredictable climates, from waterlogged soils to drought-ridden plains. It is therefore ideally suited to semi-nomadic life in Ethiopia and Eritrea, where it is farmed and exists as a staple grain. And because teff is so small, it cooks quickly, and therefore uses comparatively little fuel. Injera, the traditional Ethiopian pancake-like, slightly sour bread, is made from teff flour which is eaten not only as part of the meal but its flat surface acts as a plate, too.

A large percentage of the minute grain is composed of bran and germ, and as it's too small to process, teff is always eaten in its whole form, thus retaining all its nutrients; in fact, many of Ethiopia's famed long-distance runners attribute their energy and health to the diminutive grain. Being to all intents and purposes gluten-free (its gluten does not cause adverse reactions), teff is gaining some popularity in the health food industry as an alternative grain for those with coeliac disease, both as flour and in grain form.

Look: Tiny, poppy seed-like grains, often brown in colour.
Taste: Malty in flavour, but when fermented to make bread, it assumes a more sour taste.
Uses: Teff is most commonly ground into flour for use in breads, snacks and pancakes. The whole grain can be added to soups or boiled quickly and the crunchy grains sprinkled over vegetables or salads. It's also an ideal additional grain to add to couscous or rice and can be used in porridge, too.
Nutrients: Teff has the highest calcium content of all the grains and is an excellent source of vitamin C, a nutrient not commonly found in grains. Teff is also high in resistant starch, a type of dietary fibre which can help blood-sugar management, weight control and colon health.

Spicy avocado and quinoa salsa

We all love a chunky guacamole, but this version adds extra texture and is über good for you. Quinoa contains high levels of complete proteins, as well as plenty of vitamins and minerals, so in combination with avocado it's a superfood lover's dream! Adding a little finely chopped red onion is pleasing to some but not to others, therefore it's optional rather than essential. Serve alongside grilled fish or a griddled chicken breast, or with mounds of tortilla crisps for dipping – and always a cold bottle of beer!

Cook the quinoa in boiling water for 10–12 minutes or until tender, then drain and cool. Meanwhile, put the rest of the ingredients in a bowl with half the lime juice. Season well with salt and pepper and then stir in the quinoa. Taste for seasoning and add extra lime juice or salt and pepper as required.

Alternative to cooked quinoa: cooked couscous

_____

Serves 6 as a dip or 4 as a side dish

50g quinoa, thoroughly washed
2 avocados, chopped
1 large ripe tomato, deseeded and finely chopped
a handful of coriander, chopped
2 tablepoons extra-virgin olive oil
a pinch of sugar
½ –1 red chilli, deseeded and chopped (depending on how fiery you like it)
½ small red onion, chopped (optional)
juice of 2 limes
sea salt and freshly ground black pepper

Babaganoush with teff

Teff adds a malty richness to this Mediterranean aubergine dip. It's full of nutrients – packed with iron, thiamine, calcium, vitamin C and dietary fibre – so it's quite a wonder grain! If you can't find teff, this recipe works well with some cooked quinoa stirred through. Serve this dip with warm pitta or crudités.

Preheat the oven to 170°C/150°C fan/gas mark 3. Place the whole aubergine onto a baking tray and bake for 40 minutes to 1 hour or until soft. Meanwhile, boil the teff in water for 15 minutes, and then drain in a fine sieve.

When the aubergine is ready, remove from the oven, cool, then halve and scrape the soft flesh into a food processor. Season and add the cooked and cooled teff and all the remaining ingredients except the herbs and yogurt and blend until almost smooth. Taste for seasoning, adding more salt, pepper or lemon as required. Stir in the coriander, leaving a pinch for sprinkling on top.

Serves 6

1 x 500g aubergine
40g wholegrain teff
1 small garlic clove
1½ tablespoons olive oil
2 teaspoons sesame oil
2 tablespoons lemon juice
2 tablespoons low-fat Greek yogurt or soured cream
1 tablespoon chopped flat-leaf parsley or coriander
sea salt and freshly ground black pepper

Griddled polenta with spicy sweet and sour aubergine

In Italy, polenta is often left to firm up and then later sliced and fried or grilled. It is a wonderful creamy canvas for toppings such as mixed mushrooms, when in season, or this easy recipe using aubergines and tomatoes. You can make the polenta in advance and leave it to firm up in a container, then simply cut it up when you're ready to cook. You can use half a tin of chopped tomatoes, strained, in place of the fresh tomatoes.

Heat 1 tablespoon of the oil in a wok or frying pan and gently soften the onion for 15 minutes. Remove from the pan and set aside. Turn the heat up to medium, add another tablespoon of oil to the pan and stir-fry the aubergine for 10 minutes or so. Return the onions to the pan before adding the garlic, tomatoes, vinegar, sugar, capers, seasoning and 3 tablespoons of water. Cook for 15 minutes or until the tomatoes have collapsed and the aubergine is very soft.

Towards the end of the cooking time, trim and slice the polenta into 8 slices, each 1–2cm thick. Fry in a non-stick frying pan in olive oil until golden on both sides (a couple of minutes each side). Place 2 strips of polenta onto each plate and serve the aubergines on the side, with a few dressed rocket leaves and a sprinkling of toasted pine nuts.

Serves 4

2 tablespoons olive oil
½ red onion, finely sliced
250g aubergine, trimmed
  and cut into 2.5cm cubes
1 garlic clove, crushed
2 tomatoes, chopped
1 tablespoon red wine
  vinegar
2 heaped teaspoons soft light
  brown sugar
1 tablespoon capers, drained
  and chopped
1 quantity of Creamy
  Parmesan Polenta
  (see page 115)
sea salt and freshly ground
  black pepper

To serve:
rocket leaves dressed with
  lemon and olive oil
toasted pine nuts

Spiced sweetcorn and pumpkin fritters

My inspiration for this came from the many heavenly recipes that Yotam Ottolenghi has perfected for spicy fritters and cakes in his books – in particular his heavenly cauliflower fritters. This combination of sweetcorn, pumpkin and onion is very moreish and goes down incredibly well as a warm pre-supper bite to enjoy with drinks. The joy is that you can prepare the mixture up to four hours ahead of time, then cook the cakes just before friends arrive and keep them warm in the oven.

Put the corn cobs in the base of a steamer, covered with boiling water, and the pumpkin on top in the steaming compartment. Steam for about 15 minutes or until the pumpkin is soft. Leave both to cool, then run a sharp knife down the cobs to remove the kernels and set aside in a bowl.

Put the pumpkin into a separate bowl and mash until smooth. Beat in the eggs then add the cumin, ground coriander, salt, pepper and flour and combine well. Stir in the milk, followed by the sweetcorn kernels, spring onions, fresh coriander and chilli.

Shallow fry small spoonfuls of the mixture until golden on both sides – they take 2–3 minutes per side. Drain on kitchen paper and keep warm.

Meanwhile, mix the soured cream or yogurt, lime juice, cayenne and coriander in a bowl with some salt and pepper. Serve with the fritters.

Makes 20–30 fritters

2 large corn cobs
400g pumpkin or butternut
  squash (350g once peeled),
  cut into 2.5cm cubes
2 medium free-range eggs
¾ teaspoon ground cumin
2 teaspoons ground
  coriander
1 heaped teaspoon sea salt
  flakes or ½ teaspoon
  table salt
5 tablespoons plain flour
2 tablespoons milk
4 spring onions, trimmed
  and thinly sliced
15g chopped coriander
1 red chilli, deseeded and
  finely chopped, or
  ½ teaspoon crushed
  chilli flakes
vegetable or rapeseed oil,
  for shallow frying
freshly ground black pepper

For the dipping sauce:
200ml soured cream or
  yogurt
a good squeeze of lime juice
a pinch of cayenne pepper
a good handful of coriander
  leaves, chopped

Maize/Corn
Latin name: Zea mays

Since its earliest evolution and domestication in Mexico in around 6700BC, maize (or corn) has been a primary staple in almost every world region. It was developed from a wild grass called *Teosinte*, native to Central America, whose kernels were not fused together like the husked ear of modern corn. Over thousands of years, Native Americans cultivated those plants best suited for human consumption, increasing the size of the cob and the yields of each crop. They used corn in many ways, eating the kernels whole, ground and even fermented in chicha, the alcoholic drink still popular in the Andes. *Mahiz*, the word used by Native Americans to describe corn, means 'she who sustains us'.

When the European colonists arrived in America, they quickly learnt how to cultivate maize. From there the crop spread around the world, where it now thrives in surprisingly diverse climates and is cultivated for both human consumption and animal feed. The United States is the single largest producer (corn production is as large in the US as all other grains combined in America), followed by China. Many different varieties of maize are grown and vary in colour and shades of white, yellow, orange, red and even blue and purple. Corn is consumed as whole kernels both on and off the cob, and the kernels are also processed into meal, grits, flour, oil and syrup. The latter is a cheap sweetener that crops up in everything from ketchup to snack bars and is sometimes blamed for the rise in obesity, particularly in the United States, over recent decades. Maize does not contain as much protein as wheat, but it is a good source of fibre, potassium, vitamin C and other vitamins and minerals. Recent studies have also shown corn to have very high levels of antioxidants.

Sweetcorn/ Corn on the cob/ Baby corn

The sugar-rich varieties of maize (*Zea maya saccharata*), what we know as sweet corn, are grown and harvested on the cob. Corn on the cob is an ear of sweet corn which has been picked while the endosperm is in the so-called 'milk stage', so that the kernels are still tender. The Maya ate corn as a staple food crop off the cob, either roasted or boiled. Their surprising sweetness comes from a genetic defect – the sugars contained in the kernels do not all turn to starch – especially if you are lucky enough to eat the cobs when freshly picked. Baby corn is when the ear is harvested while immature so it is tender enough to eat whole. Although it has recently become more popular in Europe and the US, baby corn has found most widespread use in Asian cooking.

Look: Corn on the cob kernels are pearly and bead-like and usually sunshine yellow in colour. They are tiny and pale yellow on baby corn.
Taste: Sweet and buttery.
Uses: Sweetcorn kernels can be used in salads, soups and relishes. Corn on the cob is delicious barbecued, while baby corn can be served raw or cooked, in stir-fries and salads. In Latin America, sweet corn is traditionally eaten with beans; each of these foods is deficient in an essential amino acid that happens to be abundant in the other, so together sweet corn and beans form a balanced diet.
Nutrients: Corn on the cob contains both soluble and insoluble fibre and protein, as well as antioxidants. It also contains numerous vitamins, including high levels of vitamin B1, foliate, vitamin C, phosphorus, and vitamin A.

Cornmeal/polenta

Cornmeal is a staple food used all over the world and is made by grinding the dried raw grains of field corn (as opposed to sweetcorn) to a fine, medium or coarse consistency; it also comes in different colours reflecting the type of maize used, but is most commonly white or golden yellow. Polenta is the Italian word for cornmeal and originates from the Etruscan word *puls*, used to describe grain mush. Maize came to northern Italy in the 16th century from Turkey, and to this day Italians still use the word *granturco* to describe corn. Within a hundred years of its arrival, maize was being grown and made into polenta all over the region then known as Venetia, where it brought relief from hunger to the poor. While southern Italians subsisted on pasta, northern Italians ate little more than polenta for centuries. Polenta is still a staple food in northern Italy but has shaken off its image as peasant food. It can be bought either coarse-ground ('grezza') or finely ground ('fina'). Once slowly cooked in water, milk or both, polenta becomes smooth and creamy. It can also be transferred to a container, left to cool and set and then sliced and fried or grilled.

Cornmeal, or grits, is the American term used for this type of ground maize. It was an integral part of the diet of Native Americans thousands of years ago; they ground the kernels into meal which they mixed with salt and water and then baked. Cornmeal was popular during the American Civil War because it was very cheap and versatile, and it remains a traditional staple of the southern states. Grits come in white or yellow versions, like polenta, and can be prepared in the same way, either to a soft porridge-like consistency or left to set. Soft-cooked grits are popular for breakfast, or as a side dish, sometimes seasoned with salt or sugar. Cornmeal is also the essential ingredient in johnnycakes (pancakes) and cornbread, the latter often described as a 'cornerstone' of southern cooking.

Some grits, called hominy grits, are made using the ancient practice of pre-soaking the raw maize in a lime solution, which removes the bran and germ, softens and preserves the grain, and changes the nutritional value by making the amino acids more easy to absorb. The resulting hominy can be cooked whole and then used in soups, stews and casseroles, or dried and ground to form hominy grits. Both are popular in the southern US and Latin America, particularly Mexico, where ground hominy is the basis of 'masa harina', the traditional flour used to make tortillas, tamales and other local dishes.

Look: Golden yellow or white, the texture of coarse sand, similar to semolina and ground rice. Forms a smooth, pale yellow paste when mixed with water.
Taste: Creamy, mild and sweet flavoured – a great canvas for other flavours.
Uses: Cornmeal can be used in all kinds of baked goods, including bread and cakes. For Italian-style soft polenta, mix it with butter and herbs or cheeses such as Parmesan or Gorgonzola, which makes a delicious side dish. Left to set, then sliced, fried or grilled, it can be served with a topping or as an accompaniment, and is very different to its soft counterpart. Instant or quick-cook polenta is also available for a quick fix. While not quite as delicious as the cook-from-scratch version, it is definitely worth keeping a packet in the cupboard.
Nutrients: Cornmeal is gluten-free and rich in protein, vitamins and minerals.

Mini arancini

My friend, Italian cookery writer Anna del Conte, has generously allowed me to include her wonderful recipe for fried risotto balls – known as 'arancini' and originating in Sicily. You can make them to suit your own tastes, and can add all sorts of flavours to the risotto; I love adding some meat ragout, any cheese, roast pumpkin or spinach. This easy version is great for those times when you don't have any leftover risotto, as you simply boil the risotto rice in broth, but you can use leftover risotto if you have it. As for the stuffing, Anna says: 'When my grandaughter Coco first started helping me she liked to push a pea into each arancino, then she preferred a piece of mozzarella, or a bit of sausage (cooked) or of salami. So we changed through the years.' Feel free to use just Parmesan instead of the Pecorino, as the latter is expensive and can be hard to find. Eat the arancini as they are or with a tomato sauce or chutney for dipping.

Bring the stock to the boil then add the rice and cook, with the lid off, for 15–20 minutes or until the rice is tender. Drain, reserving any stock for a soup or whatever. Tip the rice into a bowl and let it cool a little. Season generously with salt, pepper and nutmeg and taste before adding two of the eggs, the butter and the Parmesan and Pecorino. Mix very well.

With damp hands, pick up some rice and shape into generous golf ball-sized balls. Stick a few peas, a piece of mozzarella, some sausage or a combination into the middle of each ball. Spread some flour on a plate and quickly roll each ball in it.

Break the remaining two eggs into a dish and lightly beat them with a fork. Roll each ball in the egg, letting any extra egg drop back into the dish. Coat the balls with the breadcrumbs and gently pat the crumbs in with your hands. Line up all the balls on a board and place the board in the fridge for at least half an hour.

Heat the oil in a large frying pan and, when it is quite hot but not smoking, add the rice balls. Fry them for about 5 minutes, turning them over so that they fry evenly all over. Drain on kitchen paper and serve.

Makes about 24 arancini

approximately 1.2 litres well-flavoured vegetable stock
300g risotto rice
a grating of fresh nutmeg
4 free-range eggs
25g unsalted butter
25g Pecorino cheese, grated
25g Parmesan cheese, grated
plain flour, for dusting
a good handful of cooked peas, 1 x 150g mozzarella ball cut into 2cm cubes, or some small pieces of sausage, for the centre
approximately 100g dry, unflavoured breadcrumbs
100ml olive oil, for frying
sea salt and freshly ground black pepper

Mini smoked salmon coulibiac parcels

Classic *kulebyáka* is a hot Russian pie whose popularity outside Russia was quite probably helped by its appearance in French culinary writer Georges Auguste Escoffier's cookery bible, *Le Guide Culinaire*, in the 1900s. Commonly, a whole piece of fish is surrounded by a flavoursome combination of rice (or, in earlier recipes, kasha), mushroom and hard-boiled egg, to soak up the juices, then encased in either a dough or flaky pastry. This recipe is a pocket-sized version and uses ready-made puff pastry. The parcels are delicious for a light lunch or starter, served warm with dressed salad leaves. They can be prepared ahead of time. You can also make them 'sausage roll-style' as a great canapé.

Soak the rice in cold water for 10 minutes, drain, then put in a pan and cover with 125ml cold water. With a lid on, bring to the boil, then remove the lid and cook for 3–4 minutes or until the water has evaporated. Pop the lid back on and gently steam-cook for a further 4 minutes, then set aside to cool.

Meanwhile, using a slotted spoon, put the quail's eggs into boiling water and cook for exactly 2 minutes. Using the same spoon, plunge the eggs immediately into ice-cold water for 5 minutes. Peel and set to one side.

Dry the egg pan and use to heat the oil. Add the onion and fry gently for 3–4 minutes before adding the mushrooms and cooking for a further 3–4 minutes. Add the wine and cook until evaporated. Remove from the heat. In a bowl combine the rice, onion mixture, herbs, cream cheese and some salt and pepper.

Roll out the pastry so that it is as thin as possible, then cut into 12 x 11cm squares. Brush the edges with egg. Divide the rice mixture between the squares, making a pile in the middle, then top each with an egg, followed by a piece of smoked salmon on top (it doesn't matter if it's two smaller slices). Press down a little to keep the pile reasonably compact. Then, using your fingers, pull up the four corners to make a parcel, sealing the sides. Brush with beaten egg and place on a baking tray lined with greaseproof paper. Chill in the fridge for up to 3 hours if you aren't cooking them straight away.

When you are ready, preheat the oven to 220°C/200°C fan/gas mark 7. Place the baking tray on the middle shelf (there's no need to remove the greaseproof paper) and cook for 15–20 minutes or until golden brown. Serve straight away or cool, then chill for later use.

Makes 12 parcels

85g Basmati rice, rinsed until the water is clear
12 quail's eggs
2 teaspoons vegetable or sunflower oil
½ small onion, finely chopped
6 chestnut mushrooms, chopped into 1cm pieces
1 tablespoon white wine
2 teaspoons chopped dill
2 teaspoons chopped flat-leaf parsley
3 tablespoons cream cheese
375g puff pastry
1 free-range egg, beaten
120g pack smoked salmon, cut into 12 slices
sea salt and freshly ground black pepper

Sushi 'my way' – maki and California rolls

Early sushi was actually a means of preserving fish –it was wrapped in fermented rice and then when the fish was needed the rice was discarded. When sushi moved from China to Japan in around the 8th century, so-called *namanare* sushi emerged which was eaten fresh, with the fish and rice eaten together. Today, sushi is a popular fast food all around the world. I'm not convinced that a trained sushi chef would think that my method of making sushi would cut the mustard. However, we think we've done a pretty good job of working out how to make it as simple as possible to make sushi at home. This is all thanks to my friend Jodie, a sushi queen, who has passed on all her excellent top tips. I've given both extravagant and storecupboard – but equally delicious – versions as I personally don't always have sushi-grade salmon to hand!

My favourite choices of sushi filling are different to other people's, so the following are just ideas. There are many other combinations to choose from – both meat, fish and vegetarian, and you can just adapt the recipe to suit you. Omelette (*tamago* in Japanese) can be added to vegetarian fillings, or you can leave it out and keep it fresh and simple. Each filling listed is enough for 2 rolls (each making approximately 20 pieces) and the rice makes enough for 8 rolls (making 80 pieces), so you can fiddle around to suit your wishes. Serve the sushi with soy for dipping, wasabi paste and pickled ginger.

To cook the rice:

Begin by putting the rice in a colander then washing it in plenty of cold water over a bowl (so that you can see the change in cloudiness) in the sink, swirling it with your hand, until the water is completely clear. Then, transfer the drained grains to a saucepan and pour in 800ml cold water. *

With a tight-fitting lid on, preferably glass so you can see through it, bring the rice to the boil over a high heat, then turn the heat down to its lowest setting, give the rice a quick stir and replace the lid. Simmer over the gentlest heat you can for 10 minutes then take the pan off the heat, without stirring or lifting the lid, and leave covered to continue softening for 20 minutes.

Turn the rice into a non-metallic bowl (the Japanese use flat wooden bowls called *hangiri*) and stir in the vinegar. Then, cover with a damp tea towel and leave to cool completely.

Makes 80 pieces

750ml volume Japanese sushi rice (makes 1.2kg cooked rice)
60ml seasoned sushi vinegar
8 square nori seaweed sheets
1–2 teaspoons toasted sesame seeds

For the omelette:
1 free-range egg
1 teaspoon dark soy sauce
½ teaspoon caster sugar

For 2 vegetarian rolls:
2 x 1cm square 20cm long strips of peeled cucumber (or 2 x 10cm), to fit the nori sheet
⅓ avocado, cut into 2cm-wide lengths
2 x 5mm square 20cm long strips peeled carrot (or 2 x 10cm)
a squeeze of Japanese mayonnaise or other mayonnaise

For 2 tuna rolls:
½ x 185g tin tuna in brine, thoroughly drained
2 heaped tablespoons Japanese mayonaisse
a couple of pinches of chilli powder, if you like them spicy

You will also need a sushi rolling mat.

* Note: The Japanese tend to say equal volumes of rice to water. However, I have found that anyone with a slightly feisty cooker or not tight enough lid will find that suddenly the rice becomes a crackling mess. My advice is to try a little more than equal measures, as suggested, and see how you get on.

To make the omelette:

The authentic way to make a Japanese omelette for sushi is by using long cooking chopsticks (which reduce the amount of air beaten into the egg) and a small rectangular pan. The egg, once lightly beaten with the sugar and soy, should be added to the pan in batches, each time making a thin omelette which is carefully rolled using the chopsticks and kept at the end of the pan. When the next batch of egg is added, the idea is to let a little go under the existing roll, before rolling the first roll over to incorporate the next omelette (you are in essence making a larger and larger omelette roll consisting of fine tightly rolled layers). After three batches of egg you can then remove the omelette and slice it into long strips. The cheat's (my) way is to simply make a thick omelette in a small omelette pan and cut it into strips. So, the choice is now yours!

To make maki rolls:

Keep a bowl of water handy. Simply lay out a seaweed sheet on your rolling mat, shiny side down and with the faint lines through the seaweed running upwards not side ways. With wet hands, take a handful of the cold sushi rice (approximately 150g) and gradually spread and push it into the seaweed sheet so that it is evenly and tightly covered right up to the edges, but with a 2cm margin of seaweed left uncovered at the top for sealing the roll. The rice should be only 5mm thick.

Using your filling of choice, place the ingredients in a line horizontally along the rice, about 4cm from the bottom. For example, if using the crab filling, place three pieces of crab in a single line along the width, then some strips of avocado and a squeeze of mayonnaise on top.

Using the mat to help you, roll up the sushi working from the bottom to the top, gently squeezing and pushing the roll as you slowly roll it, with both hands, to ensure it is as tight as possible. Wrap in clingfilm and chill until ready to use (you can serve immediately if you like.)

To cut the sushi, unwrap the roll, place it on a board and, using a wet knife, trim the ends off before cutting into approximately 10 slices (have a cup of hot water handy for knife dipping between slices). Arrange on a platter and serve with small bowls of the wasabi paste, ginger and soy.

To make California rolls

Cover the nori sheet entirely with the rice, leaving no margin at the top. Place a layer of clingfilm over the top, and ideally a second rolling mat (or a baking mat if you don't have a second mat), then flip the seaweed sheet over onto the rolling mat, so that the clingfilm is on the underside and the seaweed is facing upwards. Place the filling (any of the above) on the seaweed as for the maki and roll up so that the rice is on the outside). Roll in toasted sesame seeds, then wrap and chill. Cut as above. Arrange on a platter and serve with small bowls of the wasabi paste, ginger and soy.

For 2 smoked salmon rolls:

50g pack smoked salmon
$^1/_3$ avocado, cut into 2cm-wide lengths
2 x 1cm square 20cm long strips peeled cucumber
a squeeze of Japanese mayonnaise

For 2 crab/prawn rolls:

3 crab sticks, 6 kanikama or 6 cooked large prawns, cut in half lengthways
$^1/_3$ avocado, cut into 2cm-wide lengths
a squeeze of Japanese mayonnaise

To serve:

wasabi paste, pickled ginger and light soy sauce

Nigiri with tuna or salmon

Nigiri is sushi rice topped with raw fish or cooked prawns. I've found that the prettiest way to make it is actually to hand-mould the rice and then lay over the fish slices. The key is not to use too much rice and to choose lovely thin slices of salmon – otherwise it can all be a bit heavy.

There are various methods to make these bites. I think the easiest way is to mould sausage-shaped lengths of rice on the rolling mat then, using a wet knife, cut the sausages into 20–24 smaller even-sized lengths. Mould each piece with wet hands to make a 'nigiri' shape, i.e. oval with rounded ends. Alternatively, you can pack the rice into a clingfilm-lined container, then slice the nigiri into squares using a wet knife.

Cut the fish across the grain into thin rectangles, just larger than the rice mounds. Smear the fish with a little wasabi paste, then lay the fish over the rice, wasabi paste downwards, and press gently. Arrange the nigiri on a platter and serve with small bowls of the wasabi paste, ginger and soy.

Makes 20–24 nigiri

350g cooked sushi rice, cooled and seasoned
175g sushi-grade tuna or salmon
a little wasabi paste

To serve:
wasabi paste, pickled ginger and light soy sauce

Simple sides & salads

Jamaican rice and peas – Keralan lemon rice – Mixed wild and white rice – Mexican rice – Coconut rice – Green vegetable pilav – Pilau rice – Aromatic rice with kaffir lime, lemongrass and coriander – Freekeh pilav with fennel, pine nuts and lemon – Jewelled couscous salad – Tabbouleh – Pearl barley pilav – Quinoa, avocado and pea shoot salad – Creamy Parmesan Polenta – Kisir – Pumpkin and maceadamia nut salad with quinoa and amaranth – Mango, buckwheat and quinoa salad – Chickpea, beetroot and orange salad with freekeh

Jamaican rice and peas

Max is a great friend from London and her mum, who lives in Jamaica, has used this version of rice and peas, passed down through the family, for many years. All the way from Southfield, Jamaica, this is a simple and more importantly delicious version of the Carribean classic. Rice and peas can be served with just about any protein in the Caribbean, whether it's curried goat or jerk anything – chicken, pork, lamb, fish, shrimp, crab, you name it. Fried fish is technically meant to come with fried bammy (a.k.a. cassava bread), johnnycakes or festival dumplings, but lots of people have their fried fish with rice and peas too. It's a staple! In Jamaica they crush the spring onion whites to release the juice and add the green tops whole, but as we like spring onion in our house, I just chop them and throw them in.

Wash and soak the beans in plenty of cold water overnight, then discard the water.

The next day, add 800ml water to the beans and transfer to a medium-sized, heavy-based pot. Cover and cook on medium heat for 45–60 minutes until tender (taste the beans to check they're adequately cooked), then season with salt and pepper. Add the creamed coconut and stir until dissolved.

Wash the rice until the water runs almost clear. Add to the beans, along with the onion, spring onions, butter, thyme, Scotch bonnet, 1½ teaspoons salt and 1½ teaspoons pepper. Add enough water so that there is about 5cm water covering the mixture.

Bring to the boil over a high heat, with the lid off, then cover and simmer on low for around 30 minutes, until the water is fully absorbed. If the contents are not adequately cooked after that, make a well in the middle with the handle of a wooden spoon, add about 4 tablespoonfuls of water, cover again and cook on a low heat. Repeat the procedure until cooked. (Ensure the Scotch bonnet is not disturbed – it's best to remove it if visible at this stage. If not, remove it before you serve the rice; use a spoon, as the pepper will disintegrate.) Taste the rice, adding some extra seasoning if required, then turn the rice into a warmed dish, ready to serve.

200g dried red kidney beans
100g piece creamed coconut, chopped into fine pieces
500g Basmati rice
1 medium onion, finely chopped
4 spring onions, chopped
1 tablespoon butter
3 stalks of thyme
1 medium Scotch bonnet pepper, whole (optional, but very important in Max's mum's view)
sea salt and freshly ground black pepper

Rice

Latin name:
Oryza sativa

Rice is cultivated in over 100 countries and grown on every continent except Antarctica. Of the tens of thousands of varieties, more than 100 are commonly grown worldwide, though many of these are unfamiliar to most people. Around 96 percent of the world's rice is eaten in the area in which it is grown. The highest level of consumption, not surprisingly, is in Asia – in China, India, Indonesia, Bangladesh, Vietnam and Thailand, in that order. Unlike wheat, which is generally grown on large farms and mechanically harvested, all the rice in South and East Asia is grown on small, terraced paddy fields and harvested by hand. Cultivation methods have changed little in these parts over the centuries.

Rice is a grass and is related to the other cereals mentioned in this book, heralding from the Poaceae family of grasses. Its main species is Sativa of the genus Oryza. Its subspecies and varieties are endless, but it is simplest to concentrate on the two main subspecies of *Oryza sativa* (commonly known as Asian rice), 'Indica' and 'Japonica'. There are several characteristics that distinguish the different types of rice, but perhaps the most significant is the relative content of amylose, one of the two components of starch that affects the texture of rice once it is cooked.

Indica – long, slender-grained rice which is generally high in amylose and cooks up into fluffy grains that do not stick together (and can be eaten with fingers), though Indica does include some glutinous rice varieties. This rice is commonly grown at low elevations throughout tropical Asia, particularly in the hot equatorial climate of India; indeed most of the rice produced in South Asia is Indica rice and it accounts for more than 75 percent of global trade. Basmati and Jasmine rice are both of the Indica variety.

Japonica – short- or medium-grained rice which is low in amylose and produces cooked grains with varying degrees of 'stickiness', (which are easy to eat with chopsticks). It is cultivated in temperate and mountainous regions all over the world, from China and Japan to Australia, Egypt, Italy and Spain. Another subspecies of rice, Javanica, is sometimes ranked alongside Indica and Japonica but this term is becoming less common, in favour of tropical Japonica. This medium- to long-grained and averagely sticky rice is mostly grown in tropical conditions in Indonesia but also in the southern United States.

Rice plants come in many forms, with an average life span of 3–7 months, depending on the climate and variety. The height of the stem ranges from 50cm to more than 2 metres. Cultivated species of rice are considered to be semi-aquatic annuals as substantial amounts of water are required in their cultivation. *Oryza sativa* includes both dry and wetland varieties, though most of the world's rice is the latter. Wetland rice is often cultivated in paddy fields, irrigated or flooded portions of land where shallow water helps to prevent weeds from outgrowing the rice crop. The word 'paddy' actually refers to the rice once it has been harvested and derives from the Malay word for rice plant, *padi* and is a complete, unhulled seed of rice. One grain of paddy contains one rice kernel; this is also known as rough rice. As with barley and oats but not wheat, the husks enclosing the grains are inedible and must be removed.

The history of rice cultivation

Theories about the origins of rice cultivation vary, and cookery writer Sri Owen admits in her revered book, *The Rice Book,* that it probably began 'in a thousand places at a thousand different times'. It is difficult to identify, with total certainty, whether China, India or Thailand was the original home of the rice plant – and indeed it may have been native to all.

There appears to be some consensus among historians, however, that rice was grown as far back as 5000BC. Archaeologists have discovered rice in India, which may possibly date back to around 4500BC, though the first recorded mention originates from China in 2800BC. Legendary Chinese Emperor, Shen Nung, who ruled around that time, acknowledged the importance of rice to his people by establishing annual rice ceremonies held at sowing time, at which the Emperor would scatter the first seeds. Most likely, similar ceremonies took place throughout China with local dignitaries deputising for the Emperor. Nowadays, the Chinese honour their favourite grain by specifically dedicating one of the days of the New Year festivities to it. Rice also has special status in neighbouring Japan, where it arrived from China, or perhaps the Korean peninsula, by the 6th century AD and had become a staple food. Today, rice is the soul of the nation and the whole of Japanese cuisine focuses around it.

Despite the difficulty of identifying the original birthplace of rice, we know it was introduced to Europe and the Americas by the travellers – explorers, soldiers, merchants or pilgrims – who took with them the seeds of the crops that were grown in his home or in foreign lands. While rice is a versatile crop, not all seeds transplanted successfully, however; Great Britain was never able to cultivate rice due to its adverse climatic conditions. Certain regions of Europe such as Italy and Spain, and parts of America, provided the optimum climate, giving rise to a thriving rice industry – though in each case this developed relatively late. The Ancient Greeks and Romans imported rice from India, but it was expensive and mainly used for medicinal purposes. The earliest documentation of rice cultivation in Italy dates only from the 1400s, though rice had been growing around Naples earlier than that, possibly introduced from Spain. By the 15th century, however, rice was commonly grown in the north of Italy, which is now the largest rice producer in Europe.

During the Ottoman Empire which, at its peak encompassed most of the Middle East and most of North Africa, as well as parts of Europe, there was enormous demand for rice and it had to be imported, making it a luxury. In the Middle East today, rice is part of everyday fare in the cities but is used sparingly in most rural areas. The traditional method of cooking rice in the Middle East is to make a pilaf, coating white, long-grain rice in oil before cooking in water or stock and adding flavourings and ingredients. Descriptions of this technique first appeared in Arabic books in the 13th century. The Arabic name for plainly cooked rice is *ruzz mafalfal*, meaning 'peppered rice' – this doesn't imply that the rice is flavoured with pepper but that the grains should be separate like individual peppercorns. Short grains, which have less flavour and are more glutinous, still retain a bite when cooked and are used in stuffings and puddings.

Rice was a common crop in West Africa by the end of the 17th century. While it is often said that slaves from that area first took rice across the Atlantic to America, to the Carolinas in particular, some historians believe that rice travelled to America in 1694, in a British ship bound for Madagascar which was blown off course into the harbour of Charleston, South Carolina. The story goes that local colonists helped the crew repair their ship and the captain, James Thurber, gave them some rice seed in thanks. Another tale recounts how, almost a century later, the British nearly succeeded in destroying the nascent American rice industry: during the American Revolution, they occupied the Charleston area and sent home the entire quantity of harvested rice, failing to leave any seed for the following year's crop! The American rice industry survived largely thanks to President Thomas Jefferson, who broke an Italian law by smuggling rice seed out of Italy during a diplomatic mission in the late 1700s. The rice industry eventually moved from the Carolinas to the southern states surrounding the Mississippi basin, where it is concentrated to this day.

The cultural influence of rice

Rice is often associated with fertility and prosperity, and much folklore and legend surrounds the grain. In many countries, rice plays a crucial role in religion and is offered as a sign of respect to the gods. In Bali, Indonesia, where Hinduism flourishes, rice is considered a gift of the gods and its cultivation is closely linked to elaborate ritual and symbolism. It is believed that Lord Vishnu caused the Earth to give birth to rice and that Indra, king of the gods, taught people how to grow it. Dewi Sri is the revered Hindu goddess of rice. In Thailand, some people still believe that the rice angel, Phra Mae Phsot, will watch over those who don't waste rice – in the past, they would never throw away leftover rice for fear that they would be punished. In Japan, the Emperor is seen as the living embodiment of the god of the ripened rice plant and their Buddhist world view has ten categories of existence, with rice coming second only to the Emperor. And rice enjoys the patronage of its own god, Inari. The agricultural way of life in China has revolved around rice for thousands of years, influencing the social, economic, political and ideological developments of the country, to the point where traditional Chinese culture may be considered a 'rice culture'.

In some countries, such as Italy, rice is thrown like confetti over newly wedded couples to suggest that they will have a family of their own. In India, rice is the first food offered by a bride to her husband, to ensure fertility, and is the first solid food given to a baby as part of the so-called Annaprasana ceremony. Louisiana folklore suggests that the test of a true Cajun is whether he can calculate the exact quantity of gravy needed to accompany a crop of rice growing in a field!

Rice over the years has often been used in place of money. Even today, in Asia, bags of rice are traded with other villages for food or given as payment for use of equipment. Having rice on the table symbolises wealth and status. In the Philippines, a man unable to afford to eat two bowls of rice a day is said to be poor and in Thailand the phrase 'not having any rice to eat' is used to define someone who is destitute.

Case study – Wayan Mertha

I was lucky enough to spend an afternoon with Wayan Mertha while staying near Seminyak in southern Bali recently. Wayan walked and talked me around a nearby rice field (owned by a friend of his), and explained to me all about running his own family's rice farm. Life is hard and tiring for him. His wife and family live in Sukawati Village in eastern Bali, three to four hours by motorbike from Seminyak, where Wayan is now based for six days of the week. He always returns home on his day off to work on the farm.

Like most other rural farming families, Wayan's family grows rice primarily for their own consumption and to give as religious offerings. Any extra rice is used as payment to other villagers in return for produce. Any leftover rice is sold, but there isn't often any to spare. He and his family are lucky, though, as they have good supplies of rice, which they can eat at every mealtime. Like many Asians, Wayan defines someone who is 'poor' as being unable to afford three meals of rice a day – and there are many people in that category. Wayan is lucky that he has found work as a house-boy in Seminyak in order to earn extra money to take back home. He is extremely proud of what his family has achieved and that he can provide for his children so that they are healthy, happy and well educated.

While farms near towns can often employ workers to help, rural farms are mostly family run and harder work as machinery is limited and often shared or hired. It costs £2,000–£2,500 for a tractor to plough 100 square metres of land. Wayan's village also hires a rice-processing machine. For this, they give the owner 1–1.5kg for every 10kg of rice they have processed, as payment. Water is also crucial for successful farming in his village, and the Subak or 'water man' decides who can grow rice, and where. The crops are rotated with other plants in order to make sure the soil is used at its best for the different crops grown. The whole rice-growing process takes four months from planting the seeds to eating the rice, and one good harvest should feed Wayan's family for five months, so that they don't need to buy any extra.

At home, Wayan's day consists of a breakfast of strong coffee and rice, sometimes with meat or vegetables. Lunch and dinner is also mainly rice, but may well also come with sweet potatoes and corn, grown locally and sometimes served with fish or chicken and egg. The rice they eat is always white; they only grow red rice for religious ceremonies, as production takes much longer. Wayan is deeply religious, and so his routine is governed by ceremony. Every day he will take an offering of rice cake (made from sugar, grated coconut, rice and salt) to his local temple. The most important ceremonies require elaborate rice offerings that can take days to make by hand, with the villagers taking time off work to prepare the sweet offering, or 'dodol'. One offering, called 'lunkhead', requires a full day to prepare – coconut milk, rice, two types of sugar and salt are cooked and then intricately wrapped in 'klobot' or dried corn leaves. The shape is elongated and tied in a special pattern using string.

Planting and growing rice

Farming methods vary, depending on the variety of rice and the environment in which the rice is grown: 'upland rice' is cultivated mainly on hillsides, such as in Latin America; 'rain fed' rice (about a third of south Asian rice farming) is grown in shallow waters; 'irrigated' rice (common in mainland China), which are more or less independent of rainfall; and 'deep-water' rice, which can be grown in areas liable to flood, in much larger depths of water.

The ancient method

In many parts of South and East Asia, the ancient primitive methods are still practised as rice is grown on terraced paddy fields, cultivated by hand. It commonly takes approximately 1,260 litres of water to grow just 500g rice, usually up to a level of about 15cm. While the rice plant requires immense quantities of rainfall/water in its early days, this must be followed by a long and uninterrupted season of hot dry weather, so farmers must find ways to either flood the fields or drain the water from them at crucial periods.

Terraced fields are ploughed using a simple plough drawn by water buffalo to break up the soil, which is naturally fertilised. A log is then dragged over the field to smooth the earth, giving a level bed ready for the planting of the seedlings and watering. After one to two months, the rice plants are transplanted by hand to fields, which have been flooded by rain or river water. During the growing season, irrigation is maintained by dike-controlled channels or by hand watering. The fields are then drained before cutting the crop.

Mechanised rice production

In the more developed rice-growing regions of the world, such as the US and Australia, farming is now a precise science performed with specialised equipment and computers, without reliance on seasonal rains. An example can be seen in the Sacramento Valley of California, where laser-guided, land-levelling and re-circulating irrigation systems allow farmers to increase yield and reduce the amounts of water required. Fields are flooded in April and May and are then seeded by aeroplane. Gravity guides fresh water, pumped from deep wells, nearby rivers, canals or reservoirs to provide a constant water depth on the field of 5–8cm during the growing season. And, to ensure a consistent and healthy crop, fertilisers are evenly applied from the air. By September, the crop is mature and ready to be harvested. The rice is dried to remove moisture and then sent to mills for processing.

Milling

After the crop has been harvested, depending on its country of origin, it is threshed manually or by machine to loosen the hulls, at which point it becomes rough rice, or paddy. The rice is then slowly dried by warm air to reduce its moisture content to no more than 20 percent for milling. It is not uncommon in parts of Asia to see rice laid out to dry along the roadside. However, in most countries the drying of marketed rice takes place in mills. The outer husk of the rice is removed, but the bran layer left intact, which is what we call brown rice.

The rice is then cleaned and graded. If the brown rice is to be pearled (i.e the bran removed) and sold as white rice, it is then milled, an abrasive technique which removes the bran layer surrounding the rice grain. Smaller rice farms in Asia often share the use of privately owned machines with other local farms, often paying for them with a quantity of rice.

In the US and many parts of Europe, the cultivation and harvesting/processing of rice has undergone the same mechanisation as other grain crops, dramatically reducing the need for manpower.

Nutrients of rice

Rice is the single most important grain in terms of human nutrition, providing over a fifth of the calories consumed worldwide. Rice consists predominately of carbohydrate – a grain of rice contains around 80 percent starch. Rice also contains protein, with one serving providing about 9 percent of the average recommended daily amount. Rice is reasonably low in fat.

White, brown, red and black rice vary in the quantity of vitamins, minerals and antioxidants they contain. Black rice is highly nutritious and the real star of the show. Anthocyanins give it the same dark purple pigment that is found in blueberries and have strong anti-inflammatory and antioxidant properties. Black, brown and red rice contain high levels of dietary fibre and are an excellent source of manganese – a mineral that helps to digest fats and maximise the benefits of the proteins and carbohydrates we eat. Manganese also may help protect against cancer-causing free radicals.

Non-white rice is also a good source of essential B vitamins, especially thiamine and niacin. It also contains good amounts of folic acid. All rice contains minerals such as phosphorus, zinc, selenium, copper and iodine, among others. Finally, rice is gluten-free, and is one of the least allergenic of all grains.

Types of rice

Rice can be subdivided by means of a variety of characteristics:
1. Appearance by size – short-, medium- or long-grained. On the whole, long- and medium-grained rices are used for savoury recipes, with the exception of sushi and paella rice, for which the short-grained varieties favoured in sweet recipes are used.

2. Colour – rice can be distinguished by colour, with the hulled grains appearing in a rainbow of colours, from white and brown to black or red.

3. Processing method – brown rice retains its nutritious outer bran coating, while white rice is produced when the outer coating has been removed – this is sometimes known as pearled rice. Rice may also be flaked, puffed, ground, enriched/fortified (with vitamins and minerals), parboiled (this process is not, as the name suggests, to reduce cooking time, but in fact to retain nutrients that are sometimes lost in processing and cooking), and easy-cook/part-cooked.

4. Taste or texture – for example, the amylose (starch) content influences how glutinous or sticky the rice is when cooked.

5. Use – for instance, sushi, risotto, pilaf or pudding.

6. Variety – such as Basmati, Thai fragrant, Arborio, Bomba.

Long-grain rice

Basmati Rice

Basmati grows in the foothills of the Himalayas and other parts of north India, where it is considered the king of rice. In Hindi the word *basmati* translates as 'full of aroma', and the noticable fragrance of the rice and its particularly long grains set it apart from others.

White Basmati is milled and polished to remove the bran, as well as most of the germ and the aleurone layer (the layer between the bran and endosperm). Good-quality Basmati is left to mature in controlled conditions for up to 10 years: old rice cooks better and remains fluffy, with separate grains, whereas new rice becomes a little bit stickier when cooked. Basmati rice is also available as broken grains, which are less visually appealing and much cheaper. These are eaten commonly by the middle and lower classes in India, the larger grains being kept for the richer classes and for celebrations. It is not ground into rice flour as it is expensive.

White Basmati is now the most popular rice in the UK. Wholegrain, or brown Basmati is far more nutritious but higher in fat than white. In India, Basmati is also used for medicinal purposes: the liquid left after boiling rice, known as *kanji* (or congee), is a high-energy form of nourishment that is given to patients recovering from influenza, coughs or cold. The word *kanji* also refers to a thick gruel made by boiling a little rice with a lot of water.

Look: The long, slender, even-sized grains of white Basmati turn quite translucent when cooked and are silky to touch. Wholegrain or brown Basmati rice is chestnut-coloured, turning slightly paler when cooked.
Taste and Texture: It has a pure scent and a fresh, rich taste. The wholegrain version has a nuttier flavour and is slightly chewier.
Uses: Being very versatile, Basmati can be served simply as an accompaniment or made into a biryani, pilaf or stuffing, for example. It produces the best results when soaked for a short amount of time prior to cooking. Brown grains take longer to cook, but can be pretty much interchanged with white Basmati.

Jasmine Rice/Thai fragrant Rice

Grown for centuries in the mountain highlands of Thailand, Jasmine rice is still grown primarily in that country. The rice has a wonderful aroma and is named after the sweet-smelling jasmine flower of South-east Asia. You'll sometimes find the rice labelled as Thai (or Jasmine) hom mali rice, 'hom mali' meaning 'fragrant flower'.

Jasmine rice has shorter grains than Basmati and is slightly stickier when cooked. When white Jasmine rice is milled and polished, most of the germ and aleurone layer are removed as well as the bran. It can be eaten in white or brown form, though most South-east Asians prefer the white variety. Jasmine rice is generally cheaper than Basmati.

Look: Jasmine rice is praised for its whiteness and silkiness. Its delicate grains (white or brown) cling together a little more than Basmati grains once cooked, but still retain their shape.

Taste and Texture: As its name suggests, it has a wonderful aroma which is similar to that of the pandanus plant, and a delicious but delicate taste. It is softer, moister and less chewy than Basmati rice. The wholegrain version has a slightly nutty flavour.

Uses: The grains are ideally suited to eating with curries and sauces, as the slightly clinging grains soak up the liquid. They are the perfect accompaniment to Thai green, yellow and red curries, which tend to have a lot of liquid.

Thai Red Rice

Thai red rice is an ancient long-grained rice with a wonderfully intense red-brown coloured coating. It comes in a number of varieties variously called Red Jasmine, Red Hom Mali, Red Cargo as well as Thai Red rice, or variants of these names. The name 'Cargo' was coined because previously this type of rice was exported in bulk and packed into smaller bags by the importer – unlike white rice, which is usually packaged up before it is exported.

Thai red rice is similar to brown rice, in that it is unpolished, but the bran is red, purple or maroon in colour. Only the husks of the rice grains are removed during the milling process, thus all the nutrients, vitamins and minerals are left intact in the bran layer and in the germ. Semi-pearled red rice, often known as Coral Red (Jasmine) rice, is also available – these polished grains have a beautiful coral colour and are lighter and easier to digest than the wholegrain version. Red rice is rich in thiamin (vitamin B1), riboflavin (vitamin B2), fibre, iron and calcium. Thai red rice is believed to reduce cholesterol and improve circulation.

Look: The long, red-coated grains retain a deep burgundy colour when cooked; the outer bran splits open to reveal a bit of the inner white kernel.

Taste and Texture: Red rice has a wonderful aroma, a pleasant, earthy, nutty, slightly sweet flavour and a chewy texture (similar to that of wholegrain rice).

Uses: In Thailand, red rice is often eaten for breakfast because of its sweetness. With its hearty texture, it is ideal to serve with oily fish or rich meats such as duck as well as in salads.

Wehani® red rice

This is a trademarked, designer rice that was developed in the late 20th century by the Lundberg Family Farms of Richvale, California, which remains the sole producer. Developed using Basmati rice seeds, Wehani is a variant of long-grain brown rice, with the nutritious bran layer intact.

Look: It is a beautiful, brownish-red wholegrain rice, with unusually long grains similar to those of wild rice. When cooked, it resembles a red-tinted brown rice, though the grains remain more separate from each other.

Taste and Texture: The rice has a rich flavour redolent of roasted peanuts, with a slightly chewy texture.

Uses: Its full flavour and richness means that it can stand up to quite strong and earthy flavours such as mushrooms and chilli and spices. It's delicious in salads and can also be two-thirds cooked and used as a stuffing for chicken breasts.

Patna rice

This long-grained rice is grown in northern India and named after Patna, the capital of Bihar state, and has been grown and used as a staple food for thousands of years. It is processed in a similar way to white Basmati rice, but is softer and has a milder flavour; it also has a slightly higher fat content. Patna is used as everyday rice, as opposed to Basmati, which is often reserved for special occasions. During the Raj, Patna rice was much favoured by the British and at one time was the main long-grain rice sold in the UK; nowadays it is imported mainly from the US, but you can get it in Asian food shops, or online.

Look: Patna rice is milky white in colour with elongated kernels with slightly rounded edges. The grains are fluffy and dry when cooked and do not stick together.

Taste: Mildly aromatic and slightly earthy.

Uses: As well as holding its shape well as an accompaniment to curries and other dishes (in this sense, interchangeable with Basmati), Patna is often ground and mixed with lentils to be used as a coarse batters, fritters and dumplings.

American long-grain rice: Some 75 percent of the rice grown in the US is long-grain. Numerous long-grain varieties are commercially produced, primarily in Arkansas, Texas, Louisiana, Mississippi, Missouri and California, and are known generically as American long-grain rice; it often appears under the Uncle Ben's brand. Varieties include Carolina (including Carolina Gold, with a distinctive golden hull), Della and trademarked Texmati, both crosses between long-grain and Basmati. In general, these American rices are less fragrant than their Indian counterpart (some would say they are rather bland) and the grains aren't as long and slender, but they have a fluffy texture when cooked.

Dehradun: This is a high-quality type of Basmati rice grown in the foothills of the Himalayas near the city of Dehradun, capital of Uttarakhand state.

Louisiana Pecan rice: Also known as Wild Pecan rice, this is most commonly cultivated in the bayou country of southern Louisiana. It is similar to Basmati but is only partially milled so has the nutritional value of brown rice and has a rich, nutty flavour and a fragrance rather like popcorn.

Medium- and short-grain rice

Risotto Rice

Risotto rice is grown primarily in the so-called Risaie, the large paddy fields of the Po Valley in Piedmont and Lombardy. Production is highly mechanised and the rice is grown in flowing water at a constant temperature (known as a thermal blanket). The northern Italians have perfected the use of the starchy grains in their risottos, most famously perhaps in Piemonte's *risotto alla Milanese*.

The various types of risotto rice are characterised by the size and shape (i.e the length and thickness) of the grains. These range from 'comune' (the shortest) to 'semifino', 'fino' and 'superfino' (the longest). What they all have in common is the fact that they are higher in amylopectin (starch) than other types of rice. A risotto rice grain has two shells to it, and the outer shell breaks down as the rice cooks, releasing about half of the inner starch and giving the risotto its characteristic creaminess.

Arborio (named after the town of Arborio in the Po Valley) is a superfino and is the most widely used outside Italy. It is typically wider and longer than the others, however it is not as starchy and absorbs slightly less moisture.

Carnaroli is also superfino, the best type being Acquerello. Due to its very high starch content, Carnaroli tends to give a creamier end result than Arborio. The grains retain their flavour and moisture well, hold their shape and are the most forgiving if overcooked a little. Some say this variety is best for delicate risottos, using truffle for example.

Vialone Nano is the most widely used semifino rice, with smaller, rounder grains. It is the preferred rice of the Veneto region of Italy and is suited to rich and creamy risottos. It can absorb a lot of moisture, and has a high starch content like Carnaroli, producing a wonderfully creamy end result. Other lesser known varieties include Roma, Baldo, Ribe and Originario.

Look: Risotto rice grains vary from plumper, shorter grains to longer and more pointed ones. Once cooked, risotto rice is glossy and creamy.
Taste and Texture: A mild taste with a soft feel.
Uses: Mainly in risotto, of course, but also delicious used in rice pudding and can also be added to soups and stews.

Paella rice

Much of the rice grown in Spain is from the Valencia region, from where the dish originates, and paddy fields stretch for miles along the Mediterranean coast. However, the most highly regarded rice to use in this classic Spanish dish comes from the more southerly mountainous region of Murcia. Each year, a tiny amount (less than 1 percent of Spain's rice production) is grown around the small town of Calasparra. While the rice from the area is generically known as Calasparra, the local producers grow two historic varieties – Sollana (usually labelled as Calasparra) and Bomba. These are the world's only rice varieties to be DOC-certified. The grains are hand packed in small sacks of white cloth.

The niche cultivation of rice in this one small area, probably since the 14th century, is thanks to the the four rivers that cross this otherwise arid region. Aqueducts that deliver water from the mountain streams were first engineered by the Romans. The steady flow of cold mountain water means that the rice matures more slowly than its Valencian equivalent, producing kernels that are exceptionally dehydrated, particularly Bomba, the most prized and expensive paella rice. The short grains therefore absorb much more liquid and flavour than other rices, resulting in heavily swollen grains. Unlike risotto rice, Calasparra does not release so much starch and is therefore not creamy, with the grains remaining separate.

Look: Short, pearled grains, which swell generously once liquid is absorbed.
Taste and Texture: Bursting with the flavours it absorbs, and remains slightly al dente.
Uses: While used predominantly for paella-style recipes, this type of rice also works wonderfully in milky rice pudding (*arroz con leche*).

Japanese rice

White Japonica rice is the most common rice eaten in Japan, where it is known as 'Hakumai'. This short-grain rice, often with a distinctive sticky texture, has been a staple of the Japanese diet since ancient times and remains of fundamental importance to the local culture. Rice accounts for a major proportion of the farming output, with the Tohoku region being one of Japan's leading rice-growing areas. To avoid confusion, it is important to explain that Japonica rice is also grown elsewhere in the world, including in Australia, thriving in temperate climates.

There are three major groups of Japonica rice in Japan:
Uruchi Mai (sushi rice): this variety makes up most of the short-grain white rice that is grown and eaten in Japan. The grains are separated from their husks and then polished. Once cooked, they become tender and moist but retain an al dente texture; they are also slightly sticky, so it is possible to pick up a mouthful with chopsticks or use them for making sushi.
Look: Short, white grains. The cooked rice is almost pure white, with a silver shimmer, and is a little sticky.
Taste: Mild flavour with a hint of sweetness.
Uses: Nearly all Japanese cuisine uses Uruchi Mai for both rice bowl and sushi.

Hatsuga Genmai: this unpolished, brown Japonica/Uruchi Mai rice is not traditionally popular with the Japanese, but is gaining popularity due to its reported health benefits. It is traditionally served in an individual bowl with a separate bowl of miso soup. It requires much longer cooking and is chewier and nuttier than the white equivalent.
Look: Short grains that are pale brown in colour.
Taste and Texture: A slightly chewy texture with a hint of nuttiness.
Uses: Hatsuga genmai an be used for serving with Japanese dishes and soup, but not used for sushi.

Mochi Gome: this is, in fact, glutinous rice, regarded as a luxury in Japan and therefore often reserved for celebrations. Its plump, round grains are used to make rice cakes, rice crackers and cooking sake, and may also be ground into rice flour. Small, bite-sized glutinous rice cakes called 'mocha' are cooked as a special New Year recipe. Each family has a recipe which is passed down from mother to daughter; it is believed that eating them brings a long life and wealth.
Look: Round, white grains, plumper than Uruchi Mai.
Taste: Sticky and slightly sweet.
Uses: Rice cakes, crackers and ground into flour.

Glutinous rice

Asian, short-grained glutinous rice is also sometimes called sticky, waxy or sweet rice on account of its high concentration of amylopectin (a starch found in plants). Glutinous rice is grown and eaten in many Asian countries, including Thailand, Indonesia and China, where it has been cultivated for at least two millennia. It is often regarded as a luxury and is commonly eaten on special occasions. It contains more sugar and fat than ordinary rice, although both are grown and harvested under the same conditions. Unmilled sticky rice is also available and varies in colour from purple to black.

Look: Before cooking, glutinous rice grains appear chalk-like and opaque (unlike non-glutinous varieties which are almost translucent when raw).
Taste: Creamy and sweet, with a gluey texture after cooking.
Uses: Glutinous rice can be used as the basis for a baffling array of both sweet and savoury dishes. Fry with onions and spices, pound into cakes using nuts, raisins and coconut, or use to create porridge, soup, dumplings and fritters.

Black rice

Asian black rice is nicknamed 'Forbidden Rice' as it is said that in ancient China only emperors were allowed to eat it, due to its association with longevity and good health. It is grown in China, Indonesia and Thailand. Low in sugar, black rice is bursting with fibre and numerous antioxidants thought to help combat heart disease and cancer. It varies in stickiness and can therefore be used in a variety of dishes.

Look: Deep purple grains, with some varieties being high in gluten and therefore produce quite a sticky cooked rice, others less so.
Taste and Texture: Mild, nutty and earthy taste.
Uses: The more glutinous black rice is commonly used in Asian desserts and porridge but has great potential as a hearty constituent of breakfast cereals, drinks, cakes and biscuits. The less sticky varieties are used in savoury dishes.

Italian black rice: Known also, and variously, as Nero Venere ('black venus'), Black Venere or Nerone, this medium-grained rice is prized for its fragrant aroma, strong texture and flavour. It is grown in the Po Valley but originated in China and is related to 'forbidden' black rice from Asia (see above), though is slightly less glutinous. The black grains are easy to cook and make a striking and delicious accompaniment to savoury dishes, especially with fish and shellfish, hearty vegetable mixes and casseroles. If boiled, black rice takes slightly less time than wholegrain rice, and it can also be cooked risotto-style.

Japanese black rice: Similar to Chinese black rice ('Forbidden Rice') in taste and texture, Japanese black rice (or Kurogome) is grown from heirloom seeds and is becoming more and more popular in Japan. It is very expensive and is usually cooked with white rice, which it turns a deep purple.

Pudding rice: This is a generic term often used on packaging for short-grained white rice that is suitable for making traditional rice pudding. Obviously, there are many other varieties such as risotto, paella or glutinous rice, even black rice, that can also be used to make rice puddings.

Bhutanese red rice: A short/medium-grained wholegrain rice grown at high altitudes in the Paro Valley of Bhutan, this is a staple food of the Bhutanese people. It is semi-milled so retains its red colour from the remaining bran and has a nutty, earthy flavour. It is faster to cook than both other types of red rice and brown rice, and the end result tends to be pale pink, soft and mildly sticky. It is suited to recipes containing chilli, and rich meats such as duck and venison.

Calrose rice: Calrose is the name of the California rice industry's founding medium-grain rice variety (most of the rice grown in California is medium-grain); 'Cal' indicates its Californian origins, while 'rose', the term used for all medium-grained rice, indicates the structure of the grain. It holds flavour well, while the rice's soft and sticky consistency makes it suitable for sushi. It is the most recognised/abundant variety of California rice in the US.

Italian red rice: This is another Northern Italian medium-grain rice, known as 'Rosso Integrale', 'Rosso Selvaggio' or just 'Rosso'. It is an organic wholegrain rice with an aromatic scent and a sweet, nutty flavour. It can be used much like any other red or wholegrain rice and tastes delicious simply boiled in stock or water or prepared pilaf-style; it is also used as stuffing for peppers or tomatoes.

Red Patni rice: Short-grained, red Patni rice grows mainly in central and western India and is mostly consumed locally. The short, thick, yellowish grains are streaked with red and maroon. After the rice is husked, it is parboiled, to shorten the cooking time, and is then dried for storage. Similar to other types of red rice, it has a nutty aroma and a pleasant chewy texture. It goes with any curry, preferably spicy, coconut-flavoured ones. Cooked with coconut, it also makes a delicious pudding. Red rice flour is made into a thick batter for pancakes.

Red Camargue rice: This short-grained, unmilled variety of rice is cultivated in the wetlands of the Camargue region of southern France. It is a brownish-red, relatively sticky wholegrain rice, which has a nutty, earthy flavour and firm chewy texture. It is delicious when combined with duck and also oily fish such as salmon and also works well when cooked in the same way as a risotto.

Other rice products

Cheaper varieties of rice are used to make a range of products.

Flaked rice
Parboiled rice is husked, cleaned and cooked, then flattened by rollers until wafer thin, before being mechanically dried. The resulting flakes are flat and small, about 2mm long, and vary in thickness from translucent to thick, with a rough texture. They are light, with no distinct aroma and have a bland, gentle taste, often used to make hot cereals and puddings.

Puffed Rice
Puffed rice is not only eaten as a cereal and as compressed rice cakes, but it is also a popular ingredient in many sweet and savoury Asian recipes, and is known as *pori*. It originated, however, in the United States, thanks to an experiment performed on starch crystals by a botanist called Dr Alexander Pierce Anderson in 1901. He introduced his new puffed rice to the public in 1904 at the St Louis World's Fair in Missouri, by shooting the grains out of a gun! (The puffed grain was later nicknamed the 'Food Shot From Guns'). Puffed rice is made by heating moistened rice kernels in a special kind of pressure cooker until the moisture in the rice turns to steam and expands, thereby 'puffing' the rice. The puffed rice destined for cereals such as Rice Krispies is also toasted so that it retains its crispiness.

Ground rice
This is rice that has been ground to a slightly coarser texture than flour, and is also known as middling. It is very bland in flavour and is similar in look and taste to semolina. Ground rice is used in sweet recipes, such as shortbread and the sweet coconut cakes common in Asia. It is used worldwide as a starter weaning food for babies. Ground rice is also a useful gluten-free thickener of soups and stews.

Enriched/fortified rice
This is rice that has been treated by adding vitamins and minerals. Most rice processed in the West is enriched with a cocktail of nutrients.

Instant/pre-cooked rice
This is rice that has been cooked and then dehydrated. It is therefore much quicker to heat and serve as it simply requires rehydrating.

Converted/parboiled rice
Rice in the husk is soaked and then steamed before drying. The process ensures that more nutrients are retained and makes the rice grains harder and less likely to break. Most boil-in-the-bag rice is converted – Uncle Ben's is the most famous brand of trademarked converted rice.

Keralan lemon rice

This recipe originated in the Raheem Residency Hotel in Kerala. It was scribbled down for a friend, Emily, then tweaked a bit, then passed on – as all the best recipes are. It is best served with Indian-spiced fish kebabs or a Keralan curry. You can use fragrant Thai rice for this recipe but Basmati is better. I also love to stir-fry some chunks of okra first before adding the spices.

Fluff the rice with a fork so it has no lumps. Prepare all the ingredients so that you have everything assembled and ready to go – Blue Peter style!

Heat the oil in a wok or big frying pan (a wok is better). When the oil is hot, add the mustard seeds (they will pop), the curry leaves (they will frizzle), and the peanuts, chilli, lemon zest and turmeric. After a minute or so – not too long, or everything will burn – add the cooked rice and a sprinkling of salt. Use two wooden spoons to quickly toss and coat the rice in the flavours above. Add the juice of 1 lemon then taste, adding extra lemon juice or salt if required.

Serves 4–6

800g cooked Basmati rice, cooled (approximately 400g uncooked rice)
2 tablespoons vegetable oil
1 teaspoon mustard seeds (preferably yellow but black also fine)
20–25 small curry leaves
a small handful of unsalted peanuts
1 large mild red chilli, finely chopped
zest of 1 and juice of 1–2 lemons (depends on the juiciness of the lemons)
1 heaped teaspoon turmeric
sea salt flakes

Mixed wild and white rice

Because wild rice is expensive, combining it with another grain is a good option as the texture and flavour of the wild rice will still shine through. This recipe is the perfect accompaniment for rich or creamy recipes because of its slightly earthy/grassy flavour.

Soak the wild rice for 1 hour (any longer will cause the grains to break down), then place the drained rice into a pan with 700ml cold water. Cover, bring up to the boil, then immediately remove the lid and boil gently for 18–20 minutes.

Meanwhile, soak the Basmati rice in a bowl of cold water, then drain and add to the wild rice after its 25 minutes. Raise the heat to bring back up to the boil, stir once, then continue to boil both in the pan over a medium heat for a final 7–9 minutes or until the rice is just tender. Most of the water should have evaporated by the end of the cooking, but top up with a little boiling water if the grains start to dry out during this time.

Drain off any excess water, if you need to, and leave to rest with a tea towel and lid covering the pan for a final 5 minutes (off the heat).

Serves 2–3
as an accompaniment

50g wild rice, rinsed
150g Basmati rice, rinsed

Mexican rice

The roots of this dish are in rural Mexico, though it was the Spanish conquistadors who introduced rice to the Mexicans, along with the method of toasting the rice in oil and then adding liquid. If you want to prepare this recipe using brown rice, you will need to simmer it for longer. Therefore, just add a little extra hot stock as required, so that the rice doesn't dry out. Traditionally, you would serve the rice with refried beans, but you can make it a meal in itself with the addition of some chopped chorizo, chicken, peas, sweetcorn, peppers or a tin of drained black beans.

Heat the oil in a frying pan and add the onion and celery. Soften gently for 6–8 minutes before adding the garlic, cumin and chilli, followed by the rice. Stir over the heat for a minute to toast the grains, then add 550ml boiling hot stock and the tomatoes. Season with salt and pepper, bring to the boil, then simmer, stirring every so often, for 15–20 minutes, or until the rice is cooked and most of the liquid absorbed (add a little extra stock if needed).

Turn off the heat, then stir in the butter and a good squeeze of lime juice. Taste for seasoning and stir in the coriander just before serving.

Serves 4 as an accompaniment

2 tablespoons olive oil
1 onion, chopped
2 sticks of celery, chopped
2 garlic cloves, chopped
1 heaped teaspoon ground cumin
½ teaspoon hot chilli powder
200g long-grain rice, rinsed and drained
550–650ml chicken stock
½ x 400g tin tomatoes, blended
a knob of butter
a good squeeze of lime juice
1 heaped tablespoon chopped coriander
sea salt and freshly ground black pepper

Coconut rice

Considered one of the national dishes of Malaysia, where it is traditionally served for breakfast with a chilli sambal, *Nasi Lemak* is translated in Malay as 'fatty rice'; in Indonesia it is known as *Nasi Uduk*. Both versions consist of rice cooked in coconut cream. This is a versatile rice dish and can be served as an accompaniment to zesty Asian-style steamed fish or a tomato-based curry. Do not be tempted to stir the rice during cooking or it will go claggy.

Rinse the rice thoroughly until the water goes clear, then put in a saucepan and cover with the coconut milk and 100ml cold water. Add the remaining ingredients, including the pandan leaves if you want to add to the authentic flavour, cover with a lid and bring to the boil. Remove the lid and boil gently for 4–5 minutes or until the liquid has nearly all evaporated. Then lower the heat to the lowest setting and steam with a lid on for 5–6 minutes. Leave to rest for 5 minutes off the heat, then fluff up with a fork, pick out the ginger and serve.

Serves 3

200g (225ml volume) Basmati rice
200ml coconut milk
1 teaspoon sugar
½ teaspoon salt
2 slices of fresh ginger
2 pandan leaves, tied together (optional)

Green vegetable pilav

Sam and Sam Clark, of Moro fame, are big fans of North African pilavs and say in their cookbook, *Moro*, that 'a pilav shouldn't be *al dente*, but cooked through in a luxurious kind of way'. It is important to soak the rice first for a pilav to remove any excess starch as the rice grains should stay separate when cooked. This recipe is extremely versatile and the vegetables can be altered according to the seasons.

To prepare the Basmati rice for cooking, wash it three times in cold water then leave in a bowl of salted water for 10 minutes before draining.

Heat the oil and butter in a wide saucepan, add the onion and fry for about 8 minutes until softened. Add the garlic, cumin, cardamom, cinnamon, bay leaves and cayenne pepper and stir. Tip in the courgette and okra and stir for a minute before adding the thoroughly drained rice. Stir over the heat for 2–3 minutes.

Add the boiling hot stock, stir once, season, then cover with a lid. Simmer for 5 minutes before stirring in the broccoli, returning the lid and cooking for a further 5–6 minutes or until the rice is very nearly cooked. Turn off the heat and leave for 3 minutes with the lid on, then remove the cinnamon stick and bay leaves.

Stir in the almonds, dill, parsley and remaining butter. Taste for seasoning, then leave to rest with the lid on for a further 2 minutes before serving with the garlicky yogurt to spoon on top.

Serves 4 as an main course or 6 as a side dish

300g Basmati rice
1 tablespoon olive oil
50g butter, plus 25g for the final addition
2 smallish onions, sliced
1 garlic clove, crushed
scant ½ teaspoon cumin seeds
5 cardamom pods
1 cinnamon stick
2 bay leaves
2 pinches of cayenne pepper
1 courgette, cubed
175g okra, trimmed and cut into 5cm lengths
400ml vegetable stock
200g broccoli, broken into florets
2 tablespoons flaked almonds, toasted in a dry pan
2 tablespoons chopped dill
2 tablespoons chopped flat-leaf parsley
sea salt and freshly ground black pepper

To serve:
100g natural yogurt mixed with a little milk, some salt and pepper and ½ crushed garlic clove

Pilau rice

Pilaff, Pilav, Pulou, Plov, Paella, Biryani and Pilau – all variations of the same theme which is said to have originated in Persia. The 'pilaf' was, however, first recorded being served to Alexander the Great at a royal banquet in Greater Iran in the 4th century BC. The common theme of the dish is a basic method of cooking rice with onion, spices and stock. Indian cookery teacher Kurti Dayani recommends making pilau in a rice cooker. My method is stove-top, but feel free to try out her suggestion. This recipe for Indian pilau rice can be easily added to for more colour, texture and flavour – try frying sliced mushrooms with the onions, throwing in some dried fruit or adding some defrosted green peas near the end of cooking.

Wash the rice then leave it to soak in cold water for 20 minutes before draining thoroughly.

Heat the butter and oil in a pan over a low heat. Add the onions and soften gently for 10 minutes, then top with a lid for about 5 minutes, stirring a couple of times, until the onions are nicely browned but not burning. Add the garlic, ginger and all the spices. Stir for 30 seconds, then add the drained rice and gently stir in the boiling hot stock and a little salt.

Bring up to the boil, then simmer gently, uncovered, for about 10 minutes or until most of the water has evaporated. Top with a lid and steam-cook for a further 5 minutes or until the rice is soft. Taste for seasoning, adding a little more salt if needed. Then remove the whole spices, fluff the rice with a fork and serve in a bowl.

Serves 4 as an accompaniment

300g basmati rice
1 tablespoon vegetable oil
25g butter
1 onion, thinly sliced
2 garlic cloves
2.5cm piece fresh ginger, grated
2 cloves
2 bay leaves
½ cinnamon stick
3 cardamom pods, bashed
½ teaspoon cumin seeds
½ teaspoon ground turmeric
800ml chicken or vegetable stock
sea salt and freshly ground black pepper

Aromatic rice with kaffir lime, lemongrass and coriander

Thai fragrant rice, also known as Jasmine rice or Thai Hom Mali, is a long-grained variety with a wonderful aroma of pandan leaves. It is commonly pearled or white, the popular choice in Asia, but can also be bought as wholegrain. This scented rice and the warm citrus notes of the kaffir and lemongrass make a heavenly combination. I like to serve it with Thai chicken curry or some Asian-style steamed fish. If cooking larger amounts, you might need to steam the rice over the gentle heat for a little longer.

Rinse the rice twice and place in the pan with the same volume (325ml) of cold water. Add the salt (Thais wouldn't add salt but I like a little), lime leaves, lemongrass and chopped coriander stalks.

Cover with a lid and bring to the boil over a high heat. Remove the lid and boil vigorously for about 3 minutes or until the water has almost completely evaporated. Turn the heat right down to low, cover again with the lid and leave to gently steam for about 7 minutes. Turn off the heat and leave, with a lid on, for 6 minutes or so then fluff up with a fork, remove the lime and lemongrass and stir in the coriander leaves.

Serves 2–3

250g (325ml volume) Thai fragrant rice
½ teaspoon salt
6 large kaffir lime leaves, torn but kept whole
1 lemongrass stalk, bashed with a rolling pin then split (not all the way through)
1 small bunch of coriander, stalks finely chopped and leaves roughly chopped

Persian saffron rice

Saffron is king in Persia, and one of the most delicious recipes it is used for is a simple combination of Basmati rice and saffron with a crispy buttery crust (it is called *tahdig*, meaning 'bottom of the pot'). There are numerous variations, including the addition of sliced potato in the bottom of the pan, other spices or fruit. The rice is often served with meat or chicken.

Freekeh pilav with fennel, pine nuts and lemon

Pilav as we know it is likely to have originated in the 13th century or thereabouts, as it is first mentioned in Arabic cookbooks of that time. All recipes share the same cooking method of washing, then toasting grains in oil or butter and adding liquid to cook and soften, but the grains still remain separate. Recipes may then vary with the grain used (rice, green wheat, bulgur or barley to name a few), and the addition of spices, vegetables, herbs, fish and meat. This gentle green wheat pilav is a fabulous accompaniment to grilled, barbecued or steamed fish or chicken. You can also stir in some roast chicken and warm through at the end of cooking time.

Thoroughly wash the freekeh, then soak for 30 minutes in cold water before draining.

Heat the olive oil in a sauté pan or deep frying pan, add the fennel and onion and fry gently for around 10 minutes until softened. Add the butter, garlic, freekeh, cinnamon and bay leaves and turn in the hot pan over a medium heat for a couple of minutes.

Add the wine and lemon zest. Once the wine has evaporated, add the boiling hot stock and some seasoning and cook for 25 minutes, stirring every so often. Add the courgette and cook for a further 15–20 minutes or until nearly all the liquid has evaporated before adding the lemon juice, pine nuts and parsley. Check for seasoning and serve.

Alternative to freekeh: use bulgar wheat, which can just be rinsed thoroughly rather than soaked. Add at the same point as the freekeh but use only half the quantity of stock and add the courgette at the same time as the stock. The bulgar wheat needs to cook for only 20 minutes (rather than 40–45 minutes) before finishing with the pine nuts, etc.

Serves 2–3

125g cracked freekeh
2 tablespoons olive oil
1 fennel bulb, trimmed and thinly sliced
1 onion, sliced
25g butter
1 garlic clove, chopped
½ teaspoon ground cinnamon
2 bay leaves
3 tablespoons white wine (a good splash)
zest of 1 lemon, juice of ½
700ml vegetable or chicken stock
1 large courgette, cut into 2cm cubes
1 heaped tablespoon toasted pine nuts
1 heaped tablespoon chopped flat-leaf parsley

Jewelled couscous salad

This recipe is an adaptation of the classic Persian jewelled rice, traditionally served at weddings and other celebrations, and usually containing pomegranates, herbs and nuts for colour and texture. This couscous recipe follows the same lines. In this salad, the jewels are the nuts, carrots, cucumber and pomegranate seeds. Don't worry if one element is missing – for example, you can swap the pomegranate for dried apricot, the cucumber for blanched chopped green beans or the pine nuts for flaked almonds or pistachios. Just make it colourful!

Put the couscous in a bowl, then pour over the boiling hot stock, cover with clingfilm and leave to fluff up for 5 minutes. (Alternatively, use the steaming method on page 25). Meanwhile, toast the pine nuts and sunflower or pumpkin seeds in a dry pan and leave to cool for 10 minutes.

When it's ready, add the oil and lemon juice and some seasoning to the couscous and use a fork to stir and separate. When cool, add all the remaining ingredients, stir together and serve.

_____

Alternative to couscous: cooked quinoa or cracked or bulgar or cracked wheat (all thoroughly washed)

Variation: Jewelled couscous and millet salad
For this variation on the above recipe, you can use half the quantity of couscous and instead bring 100ml (75g) pearled millet up to a simmer in 300ml boiling stock and cook gently for 12–15 minutes or until just tender (it should have absorbed very nearly all of the liquid at the end of cooking, so watch it carefully). Mix with the other ingredients.

Serves 4 as a side dish

200g couscous
300ml chicken or vegetable stock
2 heaped tablespoons pine nuts
1 tablespoon sunflower or pumpkin seeds
2 tablespoons extra-virgin olive oil
juice of 1 lemon
a pinch or 2 of sumac (optional)
½ large cucumber, diced
1 carrot, grated
seeds from 1 pomegranate
a handful of mint leaves
a handful of flat-leaf parsley
200g feta, crumbled, (optional)

Tabbouleh

Tabbouleh is one of those recipes that everyone gets a little uptight about, with claims that theirs is the one true recipe. According to my research, everybody might be right in fact: the salad does not appear to originate from one particular nationality or culture but rather from the historic Middle Eastern region of the Levant, which encompassed a large swathe of land east of the Mediterranean Sea, which includes modern-day Israel, the Palestinian territories, Syria, Lebanon, Jordan, Cyprus and southern Turkey. So it is not surprising that the salad varies quite widely from one country to another. Some recipes I've tried are too grassy for my liking, with just a mass of herbs and not much else, while others contain way too much bulgar wheat. My favourite tabbouleh is a mildly spiced lemony version, which contains just enough wheat to add texture, some citrussy flavour and an abundance of fresh herbs. I also love to add sliced radishes and celery to mine, for crunch, and even some tuna for a lunchtime salad (horror for some I'm sure!), as in the variation below.

Wash the grains thoroughly in 2–3 changes of water. Drain, then pour 150ml boiling water over the wheat, cover with clingfilm and leave for 15 minutes. (Don't worry if there's a bit of water left behind as it will all get soaked up once the bulgar wheat's had its soak with the olive oil, etc.) Remove the clingfilm, stir in the spices, lemon juice, olive oil and some salt and pepper and leave for a further hour to soften.

Meanwhile, put the tomatoes into a small bowl and cover with boiling water. Leave for a minute, then peel off their skins, deseed and slice into slithers. Add to the bulgar wheat when it's ready, along with the cherry tomatoes, spring onions and herbs. Stir together and serve.

Variation: Crunchy tuna tabbouleh salad
To the above salad, add 1 finely sliced celery stalk, 8 thinly sliced radishes and two 185g tins of good-quality tuna in olive oil (drained). Stir together and serve.

Serves 4–6

100g bulgar or cracked wheat
2 pinches of cayenne pepper
2 pinches of ground cinnamon
2 pinches of allspice
juice of 1 lemon
3 tablespoons extra-virgin olive oil
2 tasty ripe medium tomatoes
8 cherry tomatoes, halved
3 spring onions, finely chopped
50g flat-leaf parsley leaves, chopped
50g mint leaves, chopped
sea salt and freshly ground black pepper

Pearl barley pilav

Pearl barley is a popular addition to soups and stews but can also be used in pilaf-style recipes, cooked in stock until it is just tender. This dish is wonderfully simple to prepare and is a great nutritious alternative to rice, potatoes or pasta as an accompaniment. Add some sliced mushrooms to the onions and a dash of cream at the end for a more substantial vegetarian main course.

Heat the oil in a pan. Fry the onion in the oil for 5 minutes, then add the garlic and stir for a minute. Add the rinsed and drained barley and some salt and pepper and stir for a minute before adding the boiling hot stock. Bring to the boil, cover and simmer for 25 minutes. Remove the lid and stir over the heat until the stock has nearly evaporated, without leaving the pilav dry. Taste to see if it's done, you may need a little extra liquid if needed. The grains should have a soft but chewy texture. Stir in the lemon zest and parsley and serve.

Serves 4
as an accompaniment

1 tablespoon olive oil
1 medium onion, chopped
2 garlic cloves, crushed
200g pearl barley, rinsed
1 litre chicken or vegetable
    stock
zest of 1 lemon
1½ heaped tablespoons
    chopped flat-leaf parsley
sea salt and freshly ground
    black pepper

Quinoa, avocado and pea shoot salad with tahini dressing

Some griddled halloumi or chicken would make a delicious addition to this simple salad, or serve as it is for a quick lunch or supper. It's full of ingredients that are packed with nutrients and guaranteed to make you feel good. I have given the option of adding only two-thirds of the dressing as some of those who tried it wanted it more 'wet' than others. Use chicken or vegetable stock.

Bring the stock to the boil and add the quinoa. Cook for around 12 minutes, adding the green beans for the final 2 minutes. Drain and leave to cool.

Make the dressing by simply whisking together the ingredients with half a tablespoonful of hot water and some black pepper. Put the quinoa, beans, cucumber, avocado and pea shoots (or alternatives) into a large bowl and gently toss with two-thirds of the dressing. Taste, adding extra dressing if you like. Pile the salad onto plates and serve.

Serves 2–3

100g quinoa
400ml stock
125g fine green beans
⅓–½ cucumber, halved,
    deseeded and sliced
    diagonally
1 avocado, sliced
50g pea shoots, watercress
    or rocket leaves

For the dressing:
1½ tablespoons tahini
1½ teaspoons dark soy sauce
1 garlic clove, crushed
3 tablespoons olive oil
2½ tablespoons lemon juice
½ teaspoon runny honey

Creamy parmesan polenta

This is an Italian favourite. I find that using a coarse-ground cornmeal, as I have here, produces a more sweetcorn-tasting end result for savoury dishes. However, I tend to use a finer ground polenta for sweet puddings and cakes such as the one on page 244. Well cooked polenta should be rich, creamy and glossy. Use it as a delicious alternative to mashed potato with stews and other gravy-based dishes or as a base for some roasted pumpkin and crumbled creamy blue cheese, sautéed garlic mushrooms or spring vegetables. As a starter, try the recipe on page 69.

Cooks vary in their preferred method for cooking polenta, some preferring the oven method. However, having researched and tried various methods, I am a fan of covering the polenta on top of the stove and removing the lid every few minutes to give it a stir – it produces a wonderful, creamy end result. You can use only water but I prefer a mix of milk and stock for a creamier consistency. If you like, you can also replace the coarse-ground cornmeal with the same quantity of quick-cooking instant polenta.

Pour the stock and milk into a heavy-based pan and bring up to a simmer. Whisk in the cornmeal in a steady stream and continue to whisk for 2 minutes. Then put a lid on and cook over a very gentle heat for 1 hour, stirring every 10 minutes with a wooden spoon for a good 40 seconds or so (rather like with a risotto, this breaks down the cornmeal and helps to release the starch, making it all the more creamy). If you're using quick-cook polenta, add the same quantity of liquid, but cook for a shorter time, as per pack instructions.

Stir in the butter and the Parmesan and some freshly ground black pepper and allow to stand for 3 minutes before serving.

For firm polenta, pour into a container, such as a small loaf tin lined with clingfilm, cool completely and then cut into slices before grilling or frying.

Serves 4

500ml chicken or vegetable stock
200ml milk
100g coarse-ground cornmeal, polenta or quick-cook polenta
50g butter
25g grated Parmesan
freshly ground black pepper

Kisir

Kisir is Turkey's answer to tabbouleh, with rather more grain than green plus the added flavours of pomegranate and cumin. Soaking the wheat rather than cooking it gives a nuttier bite which, in this salad, works best.

Heat 1½ tablespoonfuls of the oil in a medium-sized pan and sauté the onion, chilli and garlic for 3–4 minutes, just to take off the 'raw' edge. Stir in the cumin, then the tomato purée and cook for a minute. Add the rinsed wheat and the boiling hot stock or water, then take the pan off the heat. Cover with a lid and leave it to sit until the bulgar has cooled (about 15 minutes).

Fork the cooled grains to separate, then transfer to a bowl. Add the tomatoes, herbs and spring onions. Mix together the lemon juice, remaining olive oil and pomegranate molasses and add to the bowl along with salt and some freshly ground black pepper. Check the seasoning and serve at room temperature.

Variation: Cheat's Middle Eastern lamb burgers
Omit the dressing. To whatever weight of kisir you have, add about double the amount of raw minced lamb and some extra seasoning and combine together thoroughly. Shape into patties and grill, barbecue or fry. Serve with tzatziki and flatbreads.

Serves 4 as a side dish

3½ tablespoons extra-virgin olive oil

1 red onion, finely chopped

1 red chilli, halved, deseeded and finely shredded

2 garlic cloves, finely chopped

2 teaspoons ground cumin

2 tablespoons tomato purée

200g bulgar or cracked wheat, rinsed in 3 changes of water

250ml boiling hot vegetable stock or boiling water

4 well-flavoured plum tomatoes, peeled, deseeded and diced

a generous handful of mint leaves, torn

a small bunch of finely chopped flat-leaf parsley

3 spring onions, finely chopped

juice of 1 lemon

2 tablespoons pomegranate molasses

sea salt and freshly ground black pepper

Pumpkin and macadamia nut salad with quinoa and amaranth

You can use quinoa only for this salad (replacing 50g amaranth with extra quinoa) or, as I have done, a mix of quinoa and amaranth. I find using a potato peeler makes hard work of skinning the pumpkin – a sharp knife is easier.

Preheat the oven to 200°C/180°C fan/gas mark 6.

Put the pumpkin into a bowl with the sesame oil, olive oil, sugar and some salt and pepper and toss together. Scatter over a roasting tin, lined with greaseproof paper, in a single layer. Bake for 30–35 minutes, adding the macadamia nuts or pecans for the final 5 minutes. Leave to cool, then roughly chop the nuts.

Meanwhile, simmer the amaranth for 8 minutes in 500ml boiling water. Add the drained quinoa and boil for a further 10–12 minutes before draining and transferring to a bowl to cool slightly.

Mix all the dressing ingredients together with the finely chopped stems from the coriander. Stir into the quinoa and then add the pumpkin, nuts, onion, coriander leaves and spinach leaves. Toss lightly and serve.

———————————

Alternative to cooked quinoa: cooked freekeh or couscous or a mixture of cooked wild and basmati rice

Serves 3–4

750g pumpkin (approximately 600g peeled and deseeded), cut into 2cm-wide wedges
1 tablespoon sesame oil
1 tablespoon olive oil
½ teaspoon sugar
75g macadamia nuts or pecans
50g amaranth
100g quinoa, thoroughly washed
75g bunch of coriander, stems and leaves
1 small sweet red onion, thinly sliced
2 large handfuls of baby spinach leaves, washed
sea salt and freshly ground black pepper

For the dressing:
2 tablespoons lime juice
2 teaspoons dark soy sauce
1½ teaspoons honey
1 teaspoon sesame oil
2 teaspoons balsamic vinegar
½–1 red chilli, deseeded and finely chopped

Mango, buckwheat and quinoa salad

I love to serve this salad with big, succulent barbecued prawns or griddled chicken. Or, for a gathering, you could stir in 300g smaller, peeled, cooked prawns, adding an extra squeeze of lemon just before serving. Use a double quantity of just one of the grains if you don't have both to hand.

Bring the stock or water up to the boil and add the quinoa and buckwheat. Simmer for 10–15 minutes or until tender, then drain and cool. Meanwhile, blanch the sugar snaps in a small pan of salted water for 2 minutes, then drain and refresh in cold water, before cutting each in half. Put the sugar snaps into a large bowl along with the chopped pepper, mango, cucumber and spring onion.

Whisk together the vinegar, lemon juice and zest, ginger, olive oil, parsley, mint and a pinch or two of sugar and some salt and pepper. Put the grains into a large bowl, pour over the dressing and gently toss together.

Alternative grain: couscous or rice

Serves 6

800ml chicken or vegetable stock or water
200g red, white or rainbow quinoa, or a mixture
200g buckwheat
150g sugar snap peas, de-strung
1 red pepper, finely chopped
1 large mango, finely chopped
½ large cucumber, finely chopped
3 spring onions, chopped
sea salt and freshly ground black pepper

For the dressing:
1 tablespoon red wine vinegar
juice and zest of 1 large lemon
5cm piece fresh ginger, grated
3 tablespoons extra-virgin olive oil
2 tablespoons chopped flat-leaf parsley
2 tablespoons chopped mint
a little sugar

Buckwheat

Latin name: Fagopyrum esculentum

Other names: buckwheat groats and kasha (the toasted version), kasza, sarrasin (France)

Despite its name, buckwheat is no relation to wheat and isn't technically a grain. It is, instead, a pseudograin distantly related to rhubarb! The plant has pretty pink and white flowers popular with bees (which produce dark, amber-coloured, strongly flavoured honey from it). The name buckwheat actually comes from the Dutch word *bochweit*, meaning 'beech wheat', which refers to the plant's beech nut-shaped fruit seeds. As a crop, buckwheat has many attractive qualities: it is quick to mature, performs well in acidic and under-fertilised soil, is pest-resistant and can be grown as a 'smother' crop, used to keep weeds at bay. As an ingredient, buckwheat adds an appealingly deep flavour and texture.

Evidence suggests that buckwheat has been a source of food for some 8,000 years. It had appeared in the Balkans by around 4000BC, but seems to have played a more significant role in China and Japan, from where it spread to Russia, Turkey and on to Europe. The two largest producers of buckwheat are China and Russia.

Buckwheat is most widely known in its ground form, but it is also found as wholegrain groats, cracked, as flakes or cereal and in numerous processed products such as pasta. Both regular and toasted wholegrain buckwheat (the latter known as kasha) are used extensively in Russian and Eastern European cooking, for pilafs and salads, while the flour is used famously in French crêpes, Russian blinis and Japanese soba noodles. But buckwheat features in many and varied cuisines. In the Valtelline region of Northern Italy, locally grown buckwheat is used to make tagliatelle pasta or 'pizzoccheri', the basis of a traditional dish featuring potatoes, cabbage and cheese; you can even find gnocchi made with buckwheat. 'Kasha varnishkes' is a traditional Jewish recipe in which kasha is combined with farfalle pasta.

Look: Buckwheat has an unmistakable, triangular appearance. The groats are naturally creamy-green in colour, becoming darker brown when toasted.
Taste: The taste is green, nutty and earthy flavoured, though the flavour can vary from packet to packet – some taste milder than others. Kasha has a stronger, roasted taste.
Uses: Whole buckwheat is delicious in salads and in combination with other pseudograins such as quinoa. Kasha is a good accompaniment to robust flavours like mushrooms, onion, tomatoes or pumpkin. Try including the flour in pancakes, baked goods or even home-made soba noodles. Buckwheat cereal can be made into porridge.
Nutrients: Buckwheat contains higher levels of zinc, copper and manganese than most other cereal grains. Packed with protein, it scores highly on the amino acid scale. It has high levels of soluble fibre, which slows down glucose absorption in the body and therefore is useful for diabetics or those trying to regulate their blood sugar levels. Rutin in buckwheat increases blood circulation and helps prevent heart disease. Containing lysine, B vitamins, calcium, phosphorus and other minerals, buckwheat can also help to strengthen the kidneys. Pure buckwheat is gluten-free, but be warned that many commercial buckwheat products will contain some wheat.

Quinoa – Kaniwa

Latin name: Chenopodium spp. (quinoa), Chenopodium pallidicaule (kañiwa)

Although quinoa (pronounced KEEN-wa) has been around for thousands of years, and was the staple diet of the Incas, only now is it being hailed as a superfood in the Western world. With a flawless nutritional profile, including a high protein content and numerous vitamins and minerals, it's no wonder that it's being raved about and is fast becoming a storecupboard essential. Quinoa is actually a seed rather than a grain, and comes from the quinoa plant, a member of the spinach family. There are many different types, only a few of which are commonly available: white quinoa seeds are the most common, but you can also find black and red quinoa, both of which have slightly more flavour and crunch. Quinoa is sometimes known as 'vegetarian caviar' for the crunchiness and translucence of the small spherical grains, which never lose their shine.

Quinoa is a Spanish spelling of the original Quechua name *kinwa*, used by the Incas. The seed was cultivated by the Incas in the Andean regions of Peru, Bolivia and Chile from around 4000BC and it became a vital part of their diet; the Incas considered quinoa a sacred plant and referred to it as the 'mother seed'. Centuries later, the crop was targeted by Spanish colonisers because of its religious (and non-Christian) importance to the local people. They actively phased out its cultivation and replaced it with corn, barley and potatoes. Quinoa survived as a grain food only among the people of Bolivia's Altiplano, where it thrives at altitudes above 1,000m. Popularity then grew again and it is now cultivated the length of the Andes, primarily in Bolivia, Peru and Ecuador. Quinoa is also being grown further afield, including in the United States, but the tastiest varieties don't do well at lower elevations or on good soil.

Look out also for quinoa's cousin, kañiwa (pronounced kah-NYEE-wah), sometimes written 'cañihua'. This dark, reddish-brown pseudograin also hails from Peru and Bolivia. It contains high levels of protein (15–19 percent) and a more complete balance of amino acids than most grains. It can be purchased online.

Look: Quinoa can be black, red or, most commonly, white. The small, disk-shaped pearlescent beads have an outer rim (rather like the planet Saturn!). Uncooked, they look a little like sesame seeds, but after cooking they swell to four times their size.
Taste and Texture: Slightly crunchy with a mild, grassy flavour, rather like couscous. Quinoa seeds have a bitter-tasting coating of saponin, which protects them from birds and insects, but it is rinsed off commercially sold seeds; even so, always rinse quinoa before use. Kañiwa doesn't have this coating.
Uses: Quinoa is a great carrier of spices and other flavours, and is wonderful in soups and stews, a popular use in the Andes. It works well in salads or cook and add to fritters and burgers or use as a coating for schnitzel, fish or chicken goujons. Quinoa flakes can be used to make porridge, cakes and biscuits.
Nutrients: Quinoa contains almost double the protein content of rice and barley and is unique in that it contains the correct quantity of all nine of the essential amino acids necessary for a healthy and balanced diet. It is gluten-free, high in minerals and vitamins E and B, contains iron and phosphorus and more calcium than milk. It is also a reasonably good source of natural oil.

Chickpea, beetroot and orange salad with freekeh

This is a lively salad with a citrus dressing – and it makes a delicious accompaniment to griddled halloumi. If you want to save time, feel free to use 2–3 beetroot from a pack of pre-cooked beetroot in its natural juices. Freekeh, toasted green wheat, adds a delicious richness to the salad, but feel free to use any grains you have in the cupboard.

Put the beetroot into a pan and cover with boiling water. Bring to the boil and cook for 1¼ hours, or until tender. Cool, peel and chop into chunks.

Meanwhile, put the rinsed freekeh into a pan with the boiling hot stock or boiling salted water and the olive oil. Bring to the boil, cover and cook for 20–25 minutes or until the freekeh is cooked, then drain thoroughly and leave to cool completely.

Peel, then segment the oranges over a bowl, catching the juices for the dressing. Put the chickpeas, spring onions and orange segments into a bowl, then mix the dressing ingredients together and pour over. Season generously, then add the cooled grains. Stir well, taste for seasoning and serve topped with the beetroot and oranges, a dribble of oil and the remaining parsley.

Alternative to freekeh: Kamut, brown rice or bulgar wheat. If using bulgar wheat or kamut, cook according to the instructions on pages 24 and 26.

Serves 4

1 large or 2 smaller beetroot
100g freekeh, washed in three changes of water
300ml stock or boiling salted water
1 tablespoon olive oil
2 oranges
1 x 400g tin chickpeas, rinsed and well drained
4 spring onions, finely chopped
sea salt and freshly ground black pepper

For the dressing:
1½ tablespoons white wine vinegar
4 tablespoons extra-virgin olive oil, plus a little for drizzling
2 teaspoons Dijon mustard
1 garlic clove, crushed
2cm piece fresh ginger, grated
scant ½ teaspoon sugar
25g bunch flat-leaf parsley, leaves removed and chopped, reserving a little for sprinkling

Substantial salads

Roast tomato and pepper salad with almonds, wholegrain rice and lentils – The ultimate superfood salad with feta and mint – Cracked wheat salad with pear, goat's cheese and pomegranate vinaigrette – Gado gado – Rosemary-roasted butternut, courgette, amaranth and barley couscous salad – Couscous with griddled vegetables and pesto dressing – Greek salad with bulgar wheat – Vietnamese beef, wild rice and papaya salad – Spelt salad with shredded paprika chicken, roast cherry tomatoes and green beans – Three rice and chicken salad with toasted almonds and grapes – Roast salmon with wild rice, lentils and peppers – Wheat berry, broad bean, rocket and chorizo salad – Parma ham, mozzarella, rocket and peach salad with buckwheat and quinoa

Roast tomato and pepper salad with almonds, wholegrain rice and lentils

Combining rice with pulses is a great way to not only add extra nutritional value to a dish but also to layer textures and flavours together. Al dente green lentils with soft and chewy brown rice is a fabulous base for any number of combinations, but this herby salad with roasted peppers looks stunning on a plate and is delicious served with grilled or barbecued chicken or, for a vegetarian option, with some rocket leaves and crumbled feta thrown in.

Preheat the oven to 200°C/180°C fan/gas mark 6.

Bring the stock to the boil and add the lentils and rice. Cook over a medium heat for 30–35 minutes or until the rice and lentils are tender (add more boiling water if the lentils begin to dry out).

Toss the red peppers and 1½ tablespoonfuls of the oil in a roasting tin and roast for 20 minutes. Add the cherry tomatoes and toss, then bake for a further 5 minutes.

Meanwhile, cook the broccoli in salted water until just tender but retaining a little crunch. Drain and cut into 5cm pieces. Toast the pumpkin seeds and almonds in a dry pan until the almonds are lightly browned.

Drain the lentils and rice and put into a bowl. Add the peppers, tomatoes, broccoli, pumpkin seeds and almonds, along with the lemon juice and zest and remaining olive oil. Season with black pepper and toss. Leave to cool for 5 minutes before stirring in the spring onions and herbs. Serve at room temperature.

Alternative to the cooked rice: cooked bulgar or cracked wheat

Serves 4

800ml vegetable stock
150g green lentils
150g wholegrain Basmati rice
2 red peppers, cut into wedges
2 tablespoons extra virgin olive oil
180g cherry tomatoes, halved
200g tenderstem broccoli, trimmed
2 tablespoons pumpkin seeds
2 tablespoons chopped blanched almonds
juice and zest of 1 lemon
2 spring onions, finely chopped
2 tablespoons chopped flat-leaf parsley leaves
2 tablespoons chopped basil
freshly ground black pepper

The ultimate superfood salad with feta and mint

Freekeh, quinoa and chia, the three wonder grains, are combined with broccoli and chickpeas to make a powerhouse of a salad and, what's more, it tastes delicious too! To reduce nutrient loss, don't steam the broccoli and sugar snaps for too long, and cook the freekeh and quinoa in the vegetable cooking water rather than stock or plain water. Add the dressing to the grains while still hot, so that they absorb all the lovely flavours.

Wash grains separately in two to three changes of water or until the water runs clear. If you are using freekeh, put it into a pan and cover with the boiling hot stock or water. Bring to the boil and simmer, lid on, for 25 minutes, adding the quinoa for the final 10–12 minutes. If you are using cracked or bulgur wheat, place it in a pan with the quinoa, cover with the boiling hot stock or water, and gently boil for 12–15 minutes. When the grains are tender, drain thoroughly in a sieve to remove as much moisture as possible.

Meanwhile, steam the broccoli for about 3 minutes and the sugar snaps for 1 minute or until they are tender but still have a crunch.

In a bowl, combine the dressing ingredients with plenty of salt and pepper, taste and add extra lemon juice if needed. Mix with the still-warm grains and chickpeas. Then add the vegetables, seeds, feta, alfalfa sprouts and mixed leaves. Toss and serve straight away.

Alternative grains: cooked bulgar wheat instead of the cooked freekeh or cracked wheat; cooked couscous or amaranth instead of the cooked quinoa

Variations:
- Omit the feta and add some shredded chicken.
- Add some sliced cooked beetroot or cherry tomatoes for colour.
- Roast some butternut squash or pumpkin, then add to the salad.
- Use lentils or cannellini beans instead of chickpeas.
- Add sunflower, linseed or sesame seeds, too.

Serves 3–4

75g freekeh, cracked wheat or bulgar wheat
350ml chicken or vegetable stock, or water
50g quinoa
200g broccoli
100g sugar snap peas
½ x 400g tin chickpeas, drained and rinsed
1 tablespoon toasted pumpkin seeds
1–2 tablespoons chia seeds, preferably ground in a mill or blender
100g feta, crumbled
a handful of alfalfa sprouts
a couple of good handfuls of mixed leaves

For the dressing:
2–3 tablespoons lemon juice
1½ tablespoons Greek yogurt
1½ tablespoons extra-virgin olive oil
1 teaspoon honey, preferably Manuka
2 tablespoons chopped mint

Amaranth

Latin name: Amaranthus hypochondriacus/ caudatus/cruentus

Other names: Kiwicha

This small but mighty seed-grain was, like quinoa and chia seeds, a staple in the diets of the Aztecs. Amaranth may have lost its place in the culinary charts, but we are in the process of rediscovering the superpowers of this protein-rich, gluten-free seed. With its broad leaves and crimson flowers, amaranth is not a true grain but a member of the Chenopodiaceae family of plants and therefore related to Swiss chard and spinach, as well as to quinoa, with which amaranth has quite a lot in common. There are many cultivated species in the Amaranth genus but only three are grown for grain, the others being leaf vegetables or ornamental plants. Amaranth is fast-growing and a single plant produces dozens of seed heads, each yielding up to 5,000 seeds; these two characteristics, along with its nutritional status, have helped boost its popularity as a crop.

Amaranth originated in America and is one of the world's oldest food crops, with evidence of its cultivation dating back to the 7th millennium BC. The Aztecs and Incas believed that it had supernatural powers and incorporated it into their religious ceremonies. After the Spanish Conquest, the colonisers banned 'heathen' foods like amaranth as part of their campaign to convert the indigenous peoples to Christianity. As a result, it almost disappeared as a crop in Latin America, though it survived in some remote areas. In the ensuing centuries, amaranth cultivation spread to other parts of the world and it now grows in a wide range of habitats, including in Africa, Russia and China; there is a long history of amaranth use in India, though its origins there are unclear.

There are various forms of amaranth available, including whole seeds, flour and flakes, as well as processed products such as cereals and pasta. Cooking with amaranth can be tricky as the grains tend to form a rather gluey texture – making it ideal for porridge but not for salad – but you can mix them with a drier grain such as couscous or rice to enjoy their nutritional benefits. Mexicans pop amaranth and eat it as a sweet treat known as 'alegría' or 'happiness'.

Look: Commonly found as tiny, sand-coloured seeds, which form the same UFO-style rim as quinoa when cooked.
Taste: Mild, nutty and slightly malty.
Uses: Amaranth flour can be mixed with other flours and used in breads, muffins, crackers and pancakes as well as to thicken soups and stews. Whole amaranth can be boiled and mixed with other grains in salads and sides, or used to make porridge – the seeds won't lose their shape or crunch, but as they can become gluey they are best mixed with oats. Alternatively, try popping the seeds.
Nutrients: It is hard not to wax lyrical about the nutritious qualities of amaranth. It is valued for its protein content and contains vitamin C, rarely found in genuine grains. It has five times more iron and three times more fibre than wheat, rice or soya beans and delivers twice the amount of calcium found in milk. It is a great source of phosphorus, potassium, magnesium and manganese. High in linoleic acid, an omega-6 fatty acid, it also contains both lysine and methionine, two essential amino acids usually missing from regular grains. As if that wasn't enough, amaranth also contains phytosterols and the antioxidant rutin, which improve circulation and lower cholesterol. It is also gluten-free.

Chia

Latin name:
Salvia hispanica

Originating in Mexico, this pseudograin is one of the latest superfoods to hit our shelves, becoming increasingly well known for its impressive nutritional profile. Its status as a health food is also boosted by the fact that the tiny seeds swell to nine times their original size once eaten and therefore curb hunger pangs and snacking between meals. The seeds' satisfying, energy-boosting quality is probably why they became popular with the Aztecs as their 'running food'. It is believed that Aztec warriors insisted on drinking a mixture of chia seeds and water to give them strength in battle.

Chia, an annual summer herbaceous plant and member of the mint family, was domesticated around 2000BC. The tiny, dark seeds became one of the cornerstones of the diet of the indigenous Mayans and Aztecs, including in what is now the South-western United States. The seeds were ground into flour, pressed for oil and mixed with water as a drink. After conquering Mexico, the Spanish banned the cultivation of chia for the same reason they banned quinoa, because it played a role in non-Christians ceremonies; the seeds were also used as currency for paying tributes and taxes to the Aztec priesthood and nobility. Today, chia seeds remain a traditional staple in parts of Mexico and the southern United States. They are grown commercially primarily in Latin America.

Chia seeds can be bought whole, as flour or cold-pressed as oil. The seeds are surprisingly versatile but the classic thing to do with them is to soak them in water (or other liquid), which swells the seeds to a gel-like consistency and makes a highly nutritious drink. Note that not all the chia seeds on the market are of high quality, so it is always wise to buy the seeds (as with all grains, in fact) from a health food store that you know and trust.

Look: Tiny oval seeds with a black/brown/white base and irregular dark red-brown markings.
Taste and Texture: Raw chia seeds are hard and crunchy, with a mild taste.
Uses: Chia seeds are easy to add to just about anything – to soups or casseroles as a thickener, to muffins or health food bars, or simply as a sprinkling for your breakfast cereal. Unless you are soaking the chia in fruit juice or water for a health-giving drink, blending them in a smoothie, or adding them to porridge, a stew or soup, it is best to use cold-pressed chia or to grind the seeds a little before using, as this will prevent them passing through your body undigested – thus releasing many more nutrients into your system. Simply grind a bagful and just keep them in the fridge, ready to use.
Nutrients: Tiny wholegrain chia seeds are the world's richest known plant-based source of omega-3 fatty acids (20 percent), dietary fibre (37 percent), protein (20 percent) and antioxidants, and are burgeoning with high levels of vitamins and minerals. They are also, of course, gluten-free.

Cracked wheat salad with pear, goat's cheese and pomegranate vinaigrette

Cracked wheat is popular in Middle Eastern and Indian cookery and is used in both savoury and sweet dishes. Unlike bulgar, the wheat is not steamed or parboiled before it is cracked, and it therefore takes a little longer to cook. You can, however, use either grain for this recipe – just adjust the cooking time accordingly. This is a wonderful salad for autumn. It can be served as a starter, light lunch or on its own; it also tastes great with griddled pork steaks. Pomegranate molasses is a bottled syrup available in specialist shops and delis as well as some supermarkets.

Wash the cracked wheat in 3 changes of water or until the water runs clear. Drain thoroughly, put into a pan and cover with 600ml just boiled water. Bring to the boil, then gently boil for a further 10–12 minutes or until the wheat is tender but with a bite. Drain thoroughly, then transfer to a large bowl and leave to cool completely while you prepare the rest of the salad.

Whisk the dressing ingredients together, then add to the cracked wheat along with the pear, rocket, mint and onion in a bowl. Spoon onto a platter and sprinkle over the cheese and seeds.

Alternatives to cracked wheat: bulgar wheat, buckwheat or brown rice

Serves 6 as a side dish or 8 as a starter

150g (200ml volume) cracked wheat
2 pears, cored and chopped
100g rocket or a mix of rocket and baby leaves
handful of mint leaves, chopped
½ medium red onion, finely chopped
180g goat's or feta cheese, crumbled
seeds from ½ pomegranate
1½ tablespoons pumpkin seeds, toasted

For the dressing:
1½ tablespoons pomegranate molasses
juice of ½ lemon
3½ tablespoons extra-virgin olive oil
sea salt and freshly ground black pepper

Gado gado

In Indonesia they make this salad everywhere – the peanuts are ground and pummelled in an enormous pestle on the side of the road and the vegetables shredded at speed. Gado gado consists of a wonderful array of textures, with the firm but mild and creamy rice cake, crunchy vegetables and spicy peanut sauce. You don't have to include the *lontong* (rice cake) – some quartered boiled new potatoes, shredded chicken and/or fried tofu are also a delicious addition. For the best-tasting sauce I like to roast the peanuts until they are a rich brown, then bash them in a pestle, but for a cheat's version you can add a couple of big dollops of crunchy peanut butter to the sauce instead (in which case, reduce the sugar by half). Crispy fried shallots make a wonderful accompaniment – you can buy them ready-made in Asian supermarkets if you don't want to make your own.

For the dressing, soak the tamarind in 100ml boiling water for 10 minutes. Then, if using fresh, stone-in tamarind, scrape around the stones using the back of a spoon (discard the stones). Mix the pulp into the water.

Heat the oil in a pan. Add the onion and fry gently for a couple of minutes, then add the ginger and garlic and cook for a further minute. Add the chilli, shrimp paste (if using), sugar, tamarind and its water and the coconut cream. Roughly crush the peanuts in a pestle (or bash with a rolling pin in a bag) until crushed but not powdery. Add to the dressing along with 200ml water, bring up to a simmer and cook for 20 minutes. Loosen with a little hot water until the sauce is a pourable consistency. Cool, but keep at room temperature.

Meanwhile, blanch the cabbage for 20 seconds and the beans for 1 minute, then plunge into cold water, drain and set aside. Arrange the cabbage, beans, lontong slices, beansprouts, cucumber and eggs on a platter. Drizzle over the peanut sauce and scatter with the shallots, if you like.

Serves 6

For the dressing:
10cm whole tamarind or approximately 2 teaspoons pulped
2 teaspoons groundnut oil
½ onion, finely chopped
1 teaspoon grated fresh ginger
1 teaspoon crushed garlic
1 whole bird's eye chilli, chopped
1 teaspoon shrimp paste (optional)
2 teaspoons palm or dark brown sugar
200ml coconut cream
100g skinned peanuts, roasted

For the salad:
300g white cabbage, finely shredded
225g long or green beans, chopped into 6cm lengths
3 rolls lontong rice cake (see page 45), cut into 3cm chunks
150g beansprouts
¾ cucumber, sliced thickly
6 hard-boiled eggs, sliced into quarters

Rosemary-roasted butternut, courgette, amaranth and barley couscous salad

My father introduced me to this delicious east-west style dressing made with soy and balsamic vinegar. It's incredibly versatile and is robust enough to take on a spicy Thai-style salad as well as the rich flavours in this recipe. The barley couscous, though very tasty, is not particularly nutritious, and the amaranth, though highly nutritious, is too fine a grain to have by itself – so the combination works well!

Preheat the oven to 200°C/180°C fan/gas mark 6.

In a bowl, combine the butternut squash, rosemary, sugar, 1½ tablespoonfuls of the olive oil, a generous sprinkling of sea salt and a good grind of pepper. Scatter over a roasting tray in a single layer and roast for 20–25 minutes or until tender but not mushy. Remove from the oven and leave to cool.

Meanwhile, put the couscous into a bowl and cover with 200ml of the boiling hot stock. Cover with clingfilm and leave for 5 minutes. Put the amaranth into a pan, cover with the rest of the stock and bring to the boil. Simmer for 20 minutes, or until cooked, then pour into a sieve, rinse with a little cold water and thoroughly drain over the pan for 5 minutes.

Brush the courgette slices with the remaining oil and sprinkle with salt and pepper. Griddle until charred on both sides, then cut into 2–3cm slices.

When cooled, combine the grains in a bowl. Add the courgettes, rocket, pine nuts or pumpkin seeds, mint leaves and goat's cheese or feta (if using). Mix together the dressing ingredients and, when ready to serve, pour over the salad. Toss together, taste for seasoning and add extra lemon juice if it's not tart enough.

———————

Alternative to amaranth: cooked quinoa or an extra 50g couscous

Serves 4–6

850g butternut squash or pumpkin, peeled, deseeded and cut into 3cm chunks
2 teaspoons finely chopped rosemary leaves
½ teaspoon caster sugar
2 tablespoons extra-virgin olive oil
125g barley, wholewheat or plain couscous
700ml chicken or vegetable stock
50g amaranth
2 medium courgettes, cut lengthways into slices
50g rocket, torn
2 tablespoons toasted pine nuts or pumpkin seeds
2 heaped tablespoons chopped mint leaves
100g goat's cheese or feta, crumbled (optional)
sea salt and freshly ground black pepper

For the dressing:
2 tablespoons extra-virgin olive oil
1½ tablespoons balsamic vinegar
juice of ½ juicy lemon, plus a little extra if needed
2 teaspoons honey
1 tablespoon dark soy sauce

Couscous with griddled vegetables and pesto dressing

I make four times the quantity of pesto and freeze what I don't need in ice cube trays to use as a quick-fix pasta sauce. In this recipe, you can cook all the vegetables on the barbecue if you prefer. If you want to serve the salad cold, as an accompaniment, then just swap the halloumi for some fresh mozzarella or feta, sliced or torn. You can use a double quantity of one type of couscous if you prefer.

First, make the pesto, which you can do up to two days ahead. In a small processor or a pestle and mortar, whizz or pound the basil, garlic, lemon juice, Parmesan and pine nuts. Season generously with salt and pepper and while still processing or pounding, gradually incorporate the olive oil. Taste for seasoning and set to one side in a bowl.

Brush the vegetable slices with oil and season. Griddle in batches until nicely charred. Cut the peppers into slightly thinner strips, then set to one side. Lastly, griddle the halloumi slices on both sides.

While you're cooking the vegetables, cook the giant couscous in boiling salted water for 8–10 minutes or until tender, then drain. Place the couscous in a bowl, season and pour over 100ml boiling water and a teaspoonful of oil, then stir, cover with clingfilm and leave for 4 minutes. (Alternatively, use the steaming method on page 25.) Fork the grains, then stir in the giant couscous and half the pesto.

Toss the vegetables in the remaining pesto. Spoon the couscous onto a platter and pile the vegetables and halloumi on top.

Alternative to cooked couscous: cooked freekeh, cracked wheat or bulgar wheat

Serves 4–6

For the pesto:
40g basil, weighed with leaves and fine stalks only
½ large garlic clove, crushed
1 tablespoon lemon juice
1½ tablespoons grated Parmesan
1 heaped tablespoon pine nuts
50ml extra-virgin olive oil
sea salt and freshly ground black pepper

For the salad:
1 small aubergine, sliced
1 large courgette, sliced
1 red pepper, cut into wedges
2–3 tablespoons olive oil for brushing, plus 1 teaspoon for the couscous
180g pack halloumi, sliced
100g giant couscous
100g couscous

Greek salad with bulgar wheat

Adding bulgar wheat is a great way to pad out a traditional Greek salad and add carbs at the same time. The salad is perfect to serve on its own for lunch or with roast lamb or chicken.

Wash the bulgar wheat thoroughly in a bowl of water, then drain and wash once more. Place in a pan and cover with 400ml boiling water. Bring back to the boil then turn down the heat and simmer for about 10–12 minutes or until tender but with a nutty bite – make sure that the grains have just enough water so as not to catch. (You can also use the no-cook method of cooking bulgar described on page 24 if you prefer.) Drain thoroughly over a sieve, then cover with a lid and leave to rest for 10 minutes, before stirring with a fork and leaving to cool.

While the bulgar cools, put the red onion into a bowl with the red wine vinegar and leave to marinate. Then add the remaining ingredients along with the wheat. Season with salt and pepper, toss together and serve.

——————————

Alternative to cooked bulgar wheat: cooked cracked wheat or freekeh

Serves 4 as a main course or 6 as a side dish

100g wholegrain bulgar wheat
½ small red onion, finely sliced
2 tablespoons red wine vinegar
2 tablespoons extra-virgin olive oil
250g mixed tomatoes, cut haphazardly into chunks
1 cucumber, cut into largish chunks
1 red pepper, cut into largish chunks
a handful of mint leaves, chopped
1 Cos or Romaine lettuce, central leaves only, roughly chopped
12 pitted Kalamata olives, halved
200g pack feta cheese, drained and broken up into 3cm chunks
sea salt and freshly ground black pepper

Couscous

Other names: seksu, kesksu, taam, kuskus, kuseksi, kuskusi

Couscous is a tricky one to include because while it is grain-like in size it is, in fact, a by-product of a grain, rather than a grain in its own right. I have chosen to include it because so many people use couscous as they would other grains. Couscous is most commonly made using coarsely ground semolina made from durum wheat, but there are many different types, varying in size, shape and colour depending on the origin and grain used. Originally, couscous was made from millet and this is still used in some West African countries; other grains used to make couscous include barley and corn. To make couscous by hand is a relatively simple task but labour-intensive, involving stirring water gradually into semolina in a large bowl until the mixture forms granules; these are then dried in the sun. Nowadays, the process is largely mechanised.

Couscous is considered to be the national dish of Northwest Africa, where it has its roots with the Berber people and is served with local tagines. The first evidence of couscous is dated back to 200–150BC, and the word couscous derives from the Berber name *seksu* meaning 'well rolled' or 'rounded'. In the 11th century, the Arabo-Islamic conquest helped to distribute it east across the North African region. Economic growth and the proliferation of wheat farming then accelerated this expansion, allowing it to reach the Mediterranean.

An incredibly versatile ingredient and also quick to prepare, couscous is now used regularly in the West. The popular method of preparing it is to leave it to absorb boiling water or stock. However, in the Middle East, steaming the couscous in a couscousière or fine steamer is the preferred and more traditional method; it is often steamed directly over the simmering tagine so that it absorbs all the flavours, which I recommend doing if you have the time as the couscous will swell and fluff up even more.

Look: Small granules, similar in size to bulgar but smoother and paler coloured.
Taste and Texture: Mild and creamy, with a finer texture than rice.
Uses: Traditionally served with tagines, it is also popular as the basis for salads and side dishes and with other grains. It pairs beautifully with sunny flavours such as roasted Mediterranean vegetables, feta, nuts, fruit and herbs.
Nutrients: Plain couscous is fairly low in nutrients, compared to most grains. It contains moderate amounts of protein, carbohydrate and iron. Wholegrain couscous is better, as the semolina used retains its outer bran layer and is therefore higher in fibre. Both the plain and wholegrain varieties are low in fat.

Giant couscous

Other names: Israeli or Jerusalem couscous, pearl couscous, ptitim, fregula or fregola, cuscusu, mograbieh, miftool, mhammas, kuskus

Giant couscous is, as it sounds, a larger version of regular couscous, consisting of smooth, pearl-like balls. When cooked, it is tender but with a slight bite, tasting more like a nuttier-textured orzo pasta than regular couscous. It is also available as wholegrain. Unlike the regular version, giant couscous needs to be cooked in boiling water or stock for up to 10 minutes. It is made in a surprising number of different countries, and has a wonderful array of names as a result: fregula or fregola in Sardinia, ptitim in Israel and moghrabieh in Lebanon. You are likely to come across these names when looking for giant couscous in the shops; jumbo or Israeli couscous are other common names.

Look: Firm, bead-like granules that are pale in colour, darker if wholegrain.
Taste and Texture: Smooth, chewy, firm and slightly nutty.
Uses: Use in combination with regular couscous or on its own as a delicious addition to salads and side dishes. It's also great as a stuffing for vegetables.
Nutrients: As for regular couscous.

Wild rice
Latin name: Zizania palustris

Other names: Canada rice, Indian rice and water oats

Though its name suggests otherwise, wild rice is no relation to Asian rice (*Oryza sativa*), but is in fact the seed of a North American aquatic grass, which thrives in the cold, shallow waters of lakes, tidal rivers and bays. Evidence suggests that wild rice has been regarded as a staple food among Native American Indians for around 1,000 years. It was of prime importance in the diet of the Chippewa and Sioux Indians, native to Minnesota, where it is honoured with the title 'State Grain'. Together, Minnesota and California produce most of the wild rice grown in the United States – indeed in the world.

Cultivating wild rice, either in paddy fields or natural bodies of water, is a lot less labour intensive than collecting the truly 'wild' harvest by hand, but this is the way Native Americans still choose to do it. The indigenous peoples of North America consider the genuinely wild rice to be 'a gift from the great spirit...the creator himself' – in other words, sacred and therefore distinct from the farm-grown varieties. Collecting the wild grains is an important cultural, as well as economic, event and remains a task steeped in tradition. All but the flowering head of wild rice is below water, so the rice must be harvested from a canoe, using only a pole for power and two rice beater sticks called 'knockers' as flails to brush the mature seeds into the bottom of the boat. Wild rice is harvested green, and once the chlorophyll has dissipated from the plant, the browned rice kernel with its seed hull intact is then parched or dried to remove all moisture. It is then hulled and polished to reveal a splendid black interior.

Real wild rice, not surprisingly, is expensive. But why is the cultivated version relatively expensive, too? Mainly because it is difficult to cultivate and yields per acre are low. Producers of traditional white rice can produce a yield of 1,800 to 2,700 kilos per acre while wild rice producers can sometimes realise only 45–90 kilos of their crop per acre. To make this special treat more affordable, wild rice is often mixed with other grains, especially white and brown rice.

Look: A black, slender, rice-type grain, which opens out during cooking to reveal a pale interior.
Taste and Texture: Earthy/grassy and with a chewy texture.
Uses: It adds bite and flavour when combined with basmati rice as a side dish and is a great addition to salads. Alternatively, pop the rice in a little hot oil and eat like popcorn.
Nutrients: It is higher in protein than brown rice, and is a good source of fibre. It contains folate, magnesium, phosphorus, manganese and zinc, vitamin B6, niacin, vitamin A, omega 3, folic acid. and the amino acid lysine. It is low in fat, full of antioxidants and thought to lower cholesterol. It is also gluten-free.

Vietnamese beef, wild rice and papaya salad

Wild rice is very similar to wholegrain rice in both its texture and flavour, and it has a delicious aroma and nutty taste. It also has a similar cooking time, with the spiky hard black grains splitting when they are cooked to reveal their pale centres. Make sure not to overcook it, though – the rice should still have a pleasing bite to it. The papaya, beef and Vietnamese flavours combine with the rice in a heavenly fashion, the papaya adding a subtle sweet fruitiness to the griddled steak, along with a sharp kick from the dressing. This recipe would also work well with barbecued prawns.

Wash the wild rice and transfer to a bowl. Cover with plenty of cold water and leave to soak for 1 hour (any longer will break the grains down too much).

Drain the rice, transfer to a pan and cover with 450ml cold water. Cover with a lid, bring slowly to a boil, then remove the lid and boil gently over a medium heat for 25 minutes until the grains have just opened but still retain a little bite. Add extra boiling water if the grains are drying out. Drain off any excess water, return the rice to the pan, cover with a tea towel for 5 minutes, then fluff with a fork. Scatter the rice over a tray or large flat dish and leave to cool, then transfer to a container and chill until needed.

Meanwhile, mix the soy, garlic, sesame oil and plenty of black pepper and pour over the steaks. Thoroughly massage into the meat, then leave to marinate for 20 minutes. Preheat the barbecue or griddle pan.

Whisk all the dressing ingredients together. Place the papaya, beansprouts, cucumber, mint and rice into a large bowl.

Barbecue or griddle the steaks for 3–4 minutes on each side, or until cooked to your liking (medium rare or rare is best for this recipe). Leave the steaks to rest for 5 minutes, then slice and add to the salad, along with two thirds of the dressing. Toss together and serve immediately, adding more dressing if needed.

Serves 4 generously

100g wild rice
1 tablespoon dark soy sauce
2 garlic cloves, crushed
2 teaspoons sesame oil
3 rib-eye or fillet steaks, about 3cm thick, trimmed
1–2 firm and slightly underripe papaya (approximately 800g in weight), sliced
300g beansprouts, washed thoroughly
1 large cucumber, cut into 3mm slices on the diagonal
1 medium bunch of mint, leaves removed
freshly ground black pepper

For the dressing:
3½ tablespoons fish sauce
2 tablespoons soft brown sugar
3½ tablespoons rice vinegar
2 large red chillies, deseeded and chopped finely (keep the seeds in if you want a more fiery salad)
4 tablespoons lime juice
1 large garlic clove, crushed

Spelt salad with shredded paprika chicken, roast cherry tomatoes and green beans

This is a wonderfully easy salad that is great doubled or tripled up for larger numbers. If you like, you can cook a larger bird, to feed six – just cook the chicken until its juices run clear and up the quantities of the other ingredients by about a half again. The grains and beans can be cooked and the salad dressing made ahead of time. Use broad beans or asparagus in place of the green beans if you like.

Mix together the oil, paprika and lemon zest and juice, along with some salt and pepper, and rub all over the chicken. Leave for 20 minutes to marinate. Preheat the oven to 200°C/180°C fan/gas mark 6.

Transfer the chicken to a roasting tin and roast in the oven for 1 hour, or until the juices run clear when the thigh is pierced with a skewer. Baste the chicken once during cooking and then again just before adding the tomatoes to the tin for the final 20 minutes. At the same time as adding the tomatoes, blanch the beans in boiling salted water for 3–4 minutes, then drain and refresh the beans in cold water, before transferring to a large bowl. Put the rinsed spelt into a pan with the boiling hot stock and cook for 20 minutes or until the grains are tender but still retaining their nutty bite. Drain and add to the bowl.

Transfer the tomatoes from the roasting tin to the bowl. Put the chicken on a board and leave to rest, then spoon off any fat from the tin and add 2 tablespoonfuls of the remaining juices to the bowl. Shred the roasted chicken meat and add to the bowl along with the spring onion, cucumber, salad leaves and olives.

Whisk the dressing ingredients together. Pour two-thirds of the dressing over the salad, tossing everything together, and add more if needed.

Alternative to the spelt: pearl barley (which will need 100ml extra stock and 30 minutes cooking time)

Serves 4 generously

2 teaspoons olive oil
1 teaspoon smoked
  hot paprika
zest of 1 lemon, juice of ½
1 small chicken, weighing
  approximately 1.5kg
300g cherry tomatoes
150g fine green beans
200g spelt, rinsed
700ml chicken or
  vegetable stock
3 spring onions, finely sliced
½ cucumber, cut into
  chunks
3 handfuls of watercress,
  rocket, or baby spinach
  leaves
8 pitted Kalamata olives,
  rinsed and halved
sea salt and freshly ground
  black pepper

For the dressing:
1½ teaspoons Dijon mustard
1 teaspoon capers, drained,
  rinsed and chopped
2 tablespoons red wine
  vinegar
1 tablespoon olive oil
juice of ½ lemon
2 tablespoons chopped
  flat-leaf parsley

Three rice and chicken salad with toasted almonds and grapes

A real mix of textures and flavours, this is the perfect salad for using up leftover roast chicken or even the Christmas turkey (use brown and white meat). Add what you like – grapes can be swapped for fresh mango or orange, almonds for cashews or pecans – anything goes! A topping of crispy fried onion works well. You can roast the chicken, cook and chill the rice and prepare the dressing a few hours ahead to make life easier.

Rinse the three rices, separately, until the water runs clear. Bring a large pan of water to the boil. Add the wild rice and cook for 10 minutes before throwing in the brown rice and continuing to cook for 25 minutes. Add the white rice and cook for a further 10–15 minutes or until all are tender. Drain and cool.

In a small food processor or using a whisk, combine the dressing ingredients with some salt and pepper (if you use the processor, just roughly chop the ingredients rather than finely chop or grate). Pour half of the dressing into a large bowl and add the grapes, celery, chicken and cooled rice. Stir everything together carefully (I find clean hands are best for this task) then put into the fridge until you're ready to serve – up to 2 hours ahead is fine.

Just before serving, sprinkle in most of the almonds and the salad leaves and most of the remaining dressing. Stir carefully and taste for seasoning, adding more seasoning or dressing if needed. Top with the last few almonds and serve.

Serves 6

100g wild rice, soaked for 1 hour
100g wholegrain Basmati rice
100g white Basmati rice
250g green grapes, halved
3 sticks of celery, from the inner part, chopped
450g cooked roast chicken, cooled and cut into cubes
75g flaked almonds, toasted
50g salad leaves, such as rocket or watercress
sea salt and freshly ground black pepper

For the dressing:
1 teaspoon soft brown sugar
3 tablespoons olive oil
1 tablespoon sesame oil
½–1 red chilli, deseeded and finely chopped
3cm piece fresh ginger, grated
5 tablespoons freshly squeezed orange juice
4 tablespoons rice wine vinegar
a big handful of coriander, chopped
1 tablespoon dark soy sauce

Roast salmon with wild rice, lentils and peppers

Garlicky peppers are a perfect accompaniment to nutty wild rice and rich roast salmon with all its lovely juices. If cooking for a larger party, then simply flake the fish into the rice and everyone can dig in. If for a smaller number, serve a salmon fillet per person. You can use a mix of wild and brown rice rather than just wild rice if you like.

Preheat the oven to 200°C/180°C fan/gas mark 6. Soak the wild rice for 1 hour.

Meanwhile, deseed and cut the peppers into large wedges. Scatter onto a large baking tray, along with the thyme leaves and a good seasoning of salt and pepper. Drizzle with the oil and bake for 25 minutes.

When ready, drain the rice from the soaking water, transfer to a saucepan and cover with 1 litre of cold water. Bring slowly to the boil, then boil gently for about 30 minutes, adding the lentils after 5 minutes, and cook until both are just tender, topping up with a little water if the rice begins to dry out. Drain thoroughly and transfer to a large bowl.

Mix together the white wine and mustard with some salt and pepper and rub all over the salmon; leave to come up to room temperature. When the peppers have had their initial cooking time, remove from the oven, sprinkle over the garlic and the balsamic vinegar, give them a toss and move them up to one end of the tray. Place the salmon at the other end and bake for a final 15–20 minutes or until the salmon is just cooked through. If your baking tray is not large enough, simply cook the salmon on a separate tray, brushed with a little oil.

Using a fish slice, transfer a salmon fillet to each plate. Add the peppers and all the tray juices to the drained wild rice and lentils along with the watercress or rocket, herbs, lemon zest and juice and a good seasoning of salt and pepper and toss together. Serve the grains alongside the salmon fillets.

Serves 4

175g wild rice
150g puy or brown lentils, rinsed
1 tablespoon white wine
1 teaspoon Dijon mustard
4 x 175g thick salmon fillets
2 good handfuls of watercress or rocket
1 tablespoon chopped basil
1 tablespoon chopped flat-leaf parsley
grated zest and juice of 1 lemon
sea salt and freshly ground black pepper

For the peppers:
4 small or 3 large red peppers
4 sprigs of thyme
4 tablespoons extra-virgin olive oil
6 garlic cloves, thickly sliced
2 tablespoons balsamic vinegar

Wheat berry, broad bean, rocket and chorizo salad

You can buy wheat or rye berries from good health food shops. Wheat berries are sometimes referred to as winter wheat; they are absolutely delicious and taste a bit like a cross between pearl barley and bulgar wheat.

Wash and cook the wheat or rye berries in boiling water for 1 hour, then drain.

When the berries are nearly ready, bring a pan of salted water to the boil and throw in the asparagus and broad beans. Cook for 2 minutes, then drain. Cool slightly, then de-skin the beans and transfer to a large bowl along with the asparagus and drained berries.

Heat the oil in a frying pan over a high heat and fry the chorizo until crisp, turning it in the oil. Turn off the heat then throw in the onion, tomatoes and a good grind of pepper and toss. Add most of the lemon juice, then transfer to the bowl containing the berries.

Lastly, add the rocket leaves and toss together. Taste for seasoning, adding a little extra lemon juice if needed. Serve immediately.

––––––––––––––––

Alternative to the cooked berries: cooked couscous, bulgar wheat, quinoa or rice

Serves 4

200g wheat or rye berries
350g asparagus, woody ends removed and then cut into 7cm lengths
300g broad beans
2 tablespoons olive oil
200g chorizo, thinly sliced
1 small red onion, thinly sliced
200g cherry tomatoes, halved
juice of 1 lemon
100g rocket leaves
sea salt and freshly ground black pepper

Parma ham, mozzarella, rocket and peach salad with buckwheat and quinoa

This is a great supergrain combination, which has the added bonus of being gluten-free. Crisping the Parma ham takes less than a minute, can be done ahead of time and yet transforms an everyday salad into a wonder. Feel free to use nectarines or mango instead of peach. The salad is delicious on its own or as a tasty accompaniment to barbecued or griddled chicken.

Heat the stock or water in a pan then add the rinsed buckwheat and quinoa. Bring up to the boil and cook for 10 minutes, adding the green beans for the final 1–2 minutes, depending on the thickness of the vegetables. Drain thoroughly and leave to cool.

Meanwhile, dry-fry the Parma ham slices over a high heat for a minute or so until browning and beginning to look crisp (the ham will turn completely crispy once it has cooled). Keep to one side on a plate.

Stir the lemon juice, olive oil and some salt and pepper together in a bowl with the grains. Add the peaches, rocket, mozzarella and mint and half the Parma ham, roughly torn, and toss together. Pile onto a platter and top with the remaining crispy ham. Serve immediately.

Alternative grains: wheat berries, couscous or barley

Serves 2–3
as a main course

350ml stock or water
50g buckwheat, rinsed
50g quinoa, rinsed
100g green beans, sugar snaps or runner beans, cut into 1.5cm pieces
4 slices of very thinly sliced Parma ham
juice of ½ lemon
2 tablespoons extra-virgin olive oil
2 peaches, cut into chunks
50g rocket, torn
125g mozzarella, drained and torn
a handful of mint leaves, torn
sea salt and freshly ground black pepper

Main courses

Fragrant vegetable biryani – Spelt risotto with Mediterranean vegetables – Pourgouri with spinach and roast tomatoes – Kedgeree – Fennel and rice filo pie – Italian-style stuffed aubergines – Mexican bean and quinoa cakes – Chilli-glazed salmon and ginger parcels – Oat-coated mackerel with beetroot relish and watercress salad – Prawn, pea, lemon and mint risotto – Seafood paella – One-pot Sicilian fish stew with couscous – Salmon, dill and caper fishcakes – Thai-fried rice with prawns and pineapple – Jambalaya – Giant couscous with spicy sausages and roast vegetables – Pork fillet with prunes, mustard and barley – Chinese-style fried rice with chicken – Baked chicken, bacon and pumpkin risotto – Nasi goreng – Hainanese-style poached sesame and ginger chicken rice with greens – Roast chicken with sweetcorn stuffing – Roast duck legs with Asian-spiced red rice and plum sauce – Slow-roast lamb and bulgar wheat pilaf with saffron yogurt – Lebanese kibbeh with tahini dressing – French-style lamb and barley stew

Fragrant vegetable biryani

Biryani is an Indian spicy rice dish commonly made with mutton, lamb or chicken. Originating in the Moghul cuisine of the 16th–19th centuries, the true version is a fairly long-winded recipe, where the cooked basmati rice is piled in layers with the slow-cooked meat. It was a festive dish, costly to prepare and eaten mainly in the royal courts. Traditionally, the biryani pots were sealed with dough before baking, then cracked open at the table; this technique, called *dum*, is still used by top Indian restaurants today. This recipe is an adaptation of a stove-top version, taught to me by Indian cookery expert Kurti Dayani. It contains all the delicious spices of a biryani and is great as a vegetarian dish, simply served with raita and onion salad, or as a delicious accompaniment to tandoori chicken or curry. The choice of vegetables is best governed by the seasons as pretty much anything goes. Traditionally, this dish is sprinkled with a little rose water before serving.

Wash the rice, then leave it to soak in cold water for 20 minutes before draining thoroughly.

Heat the butter and oil in a large saucepan and soften the onions gently for 10 minutes. Then put the lid on for about 5 minutes, stirring a couple of times, until the onions are nicely browned but not burning. Add the garlic, ginger and all the spices. Stir for 30 seconds before adding the chopped tomatoes. Sauté gently for 5 minutes, stirring every so often. Add the vegetables, except for the peas and baby corn, and cook for a couple of minutes.

Meanwhile, pour enough cold water into the tomato juice jug to make the reserved tomato juice up to 800ml. Add the drained rice to the vegetables, then gently stir in the tomato juice and water and add ½ teaspoon salt. Bring up to the boil, then simmer uncovered over a medium heat for about 10 minutes or until most of the water has evaporated. Stir in the peas and corn, top with a lid and steam-cook for a further 5 minutes or until the rice is soft. You can add a little water towards the end of the cooking time if the rice starts to dry out. Taste for seasoning, adding a little more salt if needed. Then fluff the rice with a fork and serve in a bowl, sprinkled with the herbs and rose water, if using.

Serves 4

300g Basmati rice
1 tablespoon oil
25g butter
1 large or 2 small onions, chopped
2 garlic cloves, crushed
1.5cm piece fresh ginger, grated
1 x 400g tin chopped tomatoes, juice strained and reserved in a jug
½ small cauliflower, broken into florets
100g green beans, trimmed and halved
1 large carrot, cut into 2cm chunks
100g frozen peas, defrosted
6 baby corn, cut into 2cm pieces
a handful of chopped coriander and mint, for garnishing
sea salt and freshly ground black pepper

For the spices:
2 cloves
2 bay leaves
½ cinnamon stick
3 cardamom pods, bashed
1 teaspoon cumin seeds
1–2 small green chillies or 1 large, deseeded and chopped
a pinch of chilli powder
¾ teaspoon ground coriander
½ teaspoon ground turmeric

Spelt risotto with Mediterranean vegetables

This is best made in late summer when all the produce is in abundance and flavours are at their richest. Pearled spelt can be found in many health food shops; Sharpham Park is a wonderful British organic producer, whose grain products are available worldwide. Serve this risotto as it is, or as an accompaniment to sausages or roast lamb.

Heat 1½ tablespoons of the oil in a sauté pan or shallow, wide saucepan and fry the onions gently for about 10 minutes. Then transfer to a bowl and set aside. Add the courgette and pepper to the pan, season and fry for about 5 minutes over a high heat, then place in the bowl with the onions. Add another 1½ tablespoons oil and fry the aubergine over a high heat for 4–5 minutes until nearly softened, adding an extra tablespoon of oil if needed.

Turn the heat down to medium, then add the chopped tomatoes and their juice, the garlic and the rosemary. Season with salt and pepper and stir-fry for 2 minutes. Sprinkle in the spelt, stir with the tomatoes over the heat for a minute or so, then pour in the boiling hot stock and add the bay leaves. Bring up to a simmer and cook, stirring occasionally, for about 10 minutes.

Add the onion, peppers and courgettes and cook for a further 10–15 minutes or until the spelt is just tender, adding a splash of water if the risotto looks a little too dry. Squeeze over the lemon, stir and taste for seasoning, then serve.

Alternative to pearled spelt: risotto rice, wheat berries or pearl barley (may need longer cooking and extra stock)

Serves 2

3–4 tablespoons extra-virgin olive oil
2 small red onions, cut into wedges
2 small or 1 large courgettes, cut into 2.5cm chunks
1 pepper, red or orange, cut into cubes
1 x 250g aubergine, sliced and cubed
2 large ripe tomatoes, chopped
2 garlic cloves, crushed
a sprig of rosemary
100g pearled spelt, washed and drained
450ml vegetable stock
2 bay leaves
a squeeze of lemon
sea salt and freshly ground black pepper

Pourgouri with spinach and roast tomatoes

Aviva, the mother of a great friend of mine, grew up in Cyprus before marrying an Englishman, so her cooking style is a refreshing and colourful mix of influences. When talking to me of the Cypriots' love of pourgouri (bulgar wheat), she said, 'Pop into any Cypriot kitchen at lunchtime and you will probably find a casserole of pourgouri with a folded tea-towel draped between the pan and the lid. When cooked, it is rested for 10 minutes or so to absorb any steam. A Cypriot cook will tell you that your pourgouri or rice is done when little "holes" appear on the surface. Don't be tempted to lift the lid!' Serve with Greek-style yogurt or labneh. It's good as a vegetarian dish or with lamb and grilled chicken.

Wash the bulgar wheat two times and drain thoroughly. Preheat the oven to 180°C/160°C fan/gas mark 4. Mix together the olive oil, balsamic vinegar and harissa paste, season with salt, and pour over the tomatoes in a roasting tin. Turn over to coat, then leave cut-side up. Sprinkle with the sugar and roast in the oven for 45 minutes or until the tomatoes are slightly shrunken. (They can be cooked in advance.)

To make the pilaf, gently fry the chopped onion in a little olive oil in a heavy-bottomed saucepan. When translucent, add the garlic and fry for another few minutes. Add the bulgar wheat and fry gently for a minute or two, stirring every now and then so that the grains don't stick to the bottom of the pan. Take the saucepan off the heat and pour on the boiling stock. Season and stir to mix. Bring to the boil, turn down the heat and simmer, lid on, for 15 minutes until the liquid is absorbed. Remove the lid, cover with a clean tea towel, and allow to stand for 10 minutes. Fluff up with a fork.

While the pilaf rests, sauté the washed and thoroughly drained spinach in a large pan with a little olive oil, over a high heat, stirring until wilted. Squeeze out any excess moisture through a sieve, season, and stir into the pourgouri.

Quickly fry the finely sliced onions in (very) hot olive oil until golden brown with some crispy bits. Towards the end of the cooking time add the cinnamon and brown sugar. Stir around until the sugar has melted and has begun to caramelise, then add a squeeze of lemon juice and season.

Layer the pourgouri with the spinach, tomatoes and mint in a shallow bowl. Add the fried onions on top and drizzle with a little olive oil. Add small dollops of Greek yogurt or serve it separately.

For the roast tomatoes:
4 tablespoons olive oil
2 tablespoons balsamic vinegar
1–2 teaspoons harissa paste
12 medium ripe tomatoes, cut length-wise
2 teaspoons soft light brown sugar
sea salt and freshly ground black pepper

For the pilaf:
280g bulgar wheat
1 onion, finely chopped
a dash of olive oil, for frying
1 or 2 garlic cloves, finely chopped
700ml vegetable or chicken stock
400g fresh spinach leaves
extra-virgin olive oil, for drizzling
a small handful of mint leaves

For the crispy onions:
2–3 onions, very finely sliced
2 tablespoons olive oil
½ teaspoon ground cinnamon
1½ teaspoons soft brown sugar
juice of ½ lemon

To serve:
Greek yogurt

Kedgeree

This classic rice dish has rather mixed reports of its origins. Some say that kedgeree started out as an Indian rice-and-bean or rice-and-lentil dish called *khichri* (traced back to 1340 or earlier), which was then then brought back by returning British colonials who ate it as a breakfast dish in Victorian times – part of the then fashionable Anglo-Indian cuisine. Others claim that the Scots took it to India. Either way, this marvellous smoked fish and rice dish definitely hails from the days of the Raj and is still enjoyed for breakfast, lunch and supper in our house and by all generations! Some people like to use the dyed haddock for its sunshine yellow colour. However, rather than opting for an unnatural colouring, I prefer to use some turmeric in the rice instead.

Put the fish into a frying pan with the milk, bay leaves and a grinding of pepper. Bring gently up to a simmer, without letting the milk boil over. Cover with a lid and simmer for 4 minutes, then turn off the heat and set to one side while you soften the onions and spices.

Heat the oil in a large, deep frying pan or sauté pan, add the onion and soften gently for 5 minutes. Add the cardamom, cinnamon stick and curry powder and continue to cook for 5 minutes. Add the rinsed and drained rice, and stir into the onions before adding the turmeric, chopped coriander stalks, 250ml boiling water and the warm poaching milk from the fish. Bring up to nearly boiling and then simmer gently for 12–15 minutes, stirring every so often until the rice is nearly cooked through (add a little more water if the rice starts to dry out).

Meanwhile, cook the eggs in boiling water for 8 minutes, then drain and peel when cool enough to handle. Cut into wedges. De-skin and flake the fish and add to the kedgeree with the butter and chopped parsley and coriander leaves. Add a good squeeze of lemon and stir, then taste for seasoning, adding a little cayenne, salt or extra lemon juice if needed. Gently fold in the eggs and serve immediately – you can pick out the whole spices that you can see, or just warn your guests!

Serves 4

650g smoked haddock
600ml whole milk
2 bay leaves
1½ tablespoons vegetable oil
1 medium onion, chopped
8 cardamom pods, bashed
1 cinnamon stick
¾ teaspoon medium curry powder
300g Basmati rice, rinsed three times
½ rounded teaspoon turmeric
a small bunch of coriander, stalks chopped and leaves roughly chopped
3 large free-range eggs
75g butter
a handful of flat-leaf parsley, leaves removed and chopped
a squeeze of lemon
a good pinch of cayenne pepper
sea salt and freshly ground black pepper

Fennel and rice filo pie

This is one of those recipes that rather evolved over time. Years ago, when I worked for *Sainsbury's* magazine Anna del Conte did a wonderful leek, rice and Parmesan torte for a Christmas edition. I have cooked it a few times and love its simplicity. I now make it using fennel and mushrooms as well as a salmon version. It is a great entertaining recipe as it looks so beautiful and can be served warm or cold, with a salad or tenderstem broccoli, for example.

In a large, wide frying pan or wok, gently fry the fennel and mushrooms in the 25g butter and 3 tablespoonfuls of oil for 1 minute. Add the leeks and garlic and stir over the heat for a further 2–3 minutes or until just starting to soften. Add the sea salt flakes and a generous grinding of pepper, then tip in the rice, nutmeg and lemon zest. Stir for a couple of minutes to coat the rice and continue cooking the leeks, then stir in the spinach and boiling hot stock. Cook over a high heat, stirring, for 3–4 minutes or until the stock has almost evaporated. Turn off the heat and leave to cool slightly while you prepare the tin and pastry.

Preheat the oven to 200°C/180°C fan/gas mark 6. Mix the 25g melted butter and remaining 2 tablespoonfuls of oil together in a bowl. Brush a 22cm loose-bottomed, springform tin with a little of this oily mixture. Take the filo pastry and lay it out on the work surface and cover with a damp tea-towel. Line the tin with two or three sheets of the filo, so that the base and sides are covered with two layers, brushing thoroughly with the oil/butter mixture between the layers. Brush all over with more butter mixture.

Add the eggs and Parmesan to the now part-cooled rice and stir until combined. Spoon into the pastry-lined tin, then fold over the overhanging pastry to enclose the rice and then brush with butter/oil before layering a couple more sheets of the pastry on top so that the pie is completely covered. (I like to tear the sheets, then arrange them in a higgledy-piggledy fashion before brushing – it makes for a prettier top.) Brush with a little more oil/butter and place on a baking tray in the oven. Bake for 45 minutes, then remove from the oven and cool for 10 minutes in the tin, before carefully removing the springform sides and transferring to a platter. Serve in wedges.

Variation: Salmon version

Make the recipe as per the original, but fill the tin half full with the rice mixture. Then add 600g salmon fillet, skin removed and cut into 3cm thick slices, in a single layer, before topping with the remaining filling and the pastry. Use two tablespoonfuls less of grated Parmesan.

Serves 4–6

1 small (200g) fennel bulb, trimmed and thinly sliced

200g chestnut mushrooms, sliced

25g butter, plus 25g melted for brushing

3 tablespoons olive oil, plus 2 tablespoons for brushing

300g trimmed smallish leeks (approximately 4 untrimmed), white part only, thinly sliced

2 garlic cloves, crushed

¾ teaspoon sea salt flakes

200g risotto rice, such as Arborio or Carnaroli

a generous grating of nutmeg

zest of 1 lemon

200g frozen whole-leaf spinach, defrosted and water squeezed out

300ml strong vegetable or chicken stock

½ x 454g pack filo pastry

4 medium free-range eggs

50g grated Parmesan

freshly ground black pepper

Italian-style stuffed aubergines

Imam bayildi is a classic Turkish stuffed aubergine recipe featuring robust flavours of tomatoes, onion and lots of herbs and olive oil. This version has an Italian twist with the addition of creamy melted mozzarella, which goes beautifully with the wheat berries. I use this recipe as my turn-to for veggie visitors, as it needs little more than some green salad and crusty bread. If you cool the shells and filling and then stuff them and top them, they can stay in the fridge for a good few hours before baking. You can buy wheat and rye berries from good health food shops; wheat berries are sometimes referred to as winter wheat.

First, put the wheat or rye berries into a saucepan, cover with plenty of boiling water and boil for 55 minutes or until tender, topping up with water as needed. Meanwhile, preheat the oven to 200°C/180°C fan/gas mark 6.

Heat 3 tablespoonfuls of olive oil in a deep frying pan and gently fry the onion. While the onion begins to soften, run a very sharp knife around the edge of the aubergine shells, about 7.5mm in from the edge. Then cut almost through to the base in criss-cross cuts and use a teaspoon to scoop out the flesh. Roughly chop the flesh and stir into the onion. Season and cook for about 10 minutes over a medium heat until the aubergine looks half cooked. Add the garlic for the final couple of minutes.

Brush the aubergine shells with the remaining tablespoon of olive oil and season. Place on a baking tray and bake for 20 minutes. Leave to cool.

Turn down the oven to 180°C/160°C fan/gas mark 4. Add the tomato purée to the aubergines in the pan and stir for a minute before adding the chopped tomatoes and anchovies or capers. Season and add 5 tablespoons water before covering and simmering gently for 10 minutes. Stir in the drained wheat berries and basil, taste for seasoning and leave to cool. Spoon the stuffing into the shells. Top with the cheeses and bake in the oven for 15 minutes. (Alternatively you can leave everything to cool, then stuff and top with cheese before baking, adding 10 minutes to the cooking time.) Sprinkle with a little chopped basil and serve.

Alternative to cooked berries: cooked giant couscous, spelt or pearl barley

Serves 3 as a main or
6 as a side

75g wheat or rye berries
4 tablespoons olive oil
1 onion, chopped
3 x 275g aubergines (small to
  medium in size), halved
  lengthways
2 large garlic cloves, crushed
2 teaspoons tomato purée
550g (approximately 5–6
  large) tomatoes, peeled and
  roughly chopped
2 large tinned anchovy fillets,
  drained, or 1½ heaped
  teaspoons capers, chopped
12 large basil leaves,
  chopped, plus extra
  to serve
100g mozzarella, grated
3 heaped tablespoons grated
  Parmesan
1 tablespoon chopped
  basil leaves

Rye
Latin name: Secale cereale

Rye is the second most widely used cereal for breadmaking, although its gross production is less than one-fifteenth of that of wheat. It is a member of the wheat tribe, and reportedly co-evolved with wheat and barley before its value as a crop was recognised. Rye is a hardy, fast-growing crop that is able to survive more severe winters than most grains and can be grown on soils too poor for other grains. Given its hardiness, rye has long been a popular crop in colder temperate countries, including Russia and Poland (the two largest producers), Canada and China.

Modern rye is thought to have developed from either a Mediterranean variety, *Secale montanum*, or a Turkish variety, *Secale anatolicum*. The cultivation of rye took off relatively late compared with other grains, the earliest archaeological evidence dating from around 1500BC. Rye was taken to the Americas by European settlers and finally, in the 19th century, spread to South Africa and Australia. Because it grew in poor soils, rye was seen for many centuries and in many countries as the 'poverty grain' (as opposed to wheat, eaten by the rich). Not surprisingly therefore, as standards of living rose in varied civilisations, the consumption of rye declined. Yet in some food cultures, such as those of Scandinavian and Eastern European countries, rye has retained its popularity. Hopefully, as more and more people discover the cereal's nutritional benefits and its unique taste, it will assume a more important role in our diets.

Rye grain is mainly used in bread and flour, and in alcoholic drinks such as rye beer and some whiskeys and vodkas; it is also used as animal fodder. Because its gluten is less elastic than wheat's, and holds less gas during the leavening process, breads made with rye flour are compact and dense. Rye grains can also be eaten whole or as flakes.

Look: Rye berries look similar to wheat berries but are longer and thinner. Their colour varies from yellowish brown to greyish green. The flakes are a rich brown.

Taste and Texture: The berries are chewy, juicy, hearty and sweet. The flakes are rich-tasting and best used in combination with other grains.

Uses: Rye flour is used in combination with molasses to make dark and dense pumpernickel breads, but can be used in lighter breads too. You can also include rye berries in bread, stews, soups, salads, and rye flakes in porridge, muesli and granola.

Nutrients: Because it is difficult to separate the germ and bran from the endosperm, flour made from rye usually retains a large quantity of nutrients, in contrast to refined wheat flour. Rye fibre is bursting with non-cellulose polysaccharides, a class of carbohydrates that has an exceptionally high water-binding capacity and gives a feeling of fullness and satiety; this makes rye bread a great grain to consume if you are trying to lose weight. It is a good source of manganese, phosphorus and magnesium and contains lignan phytonutrients, phytochemicals thought to guard against breast and prostate cancers.

Triticale

Latin name: Triticosecale

Triticale (pronounced trih-tih-KAY-lee) is a fairly 'new' grain, created by cross-pollinating wheat with rye to produce a grain which has all the good baking qualities of wheat but also the ability, like rye, to grow in poor soils and adverse conditions, as well as a better nutritional profile.

The first wheat and rye cross-breeding took place in Scotland in 1875, but the result was sterile. In 1888, German botanists discovered how to produce a fertile hybrid and the name triticale first seems to have been used in Germany in about 1935. Although great hopes were expected for triticale, it has been slow to find widespread commercial acceptance or demand except as animal fodder. However, as it requires few pesticides, reduces soil erosion, and can capture excess soil nitrogen, triticale may well be a grain to watch in the future. Today, about 80 percent of the world's triticale is grown in Europe.

Triticale is made into flour, but can also be bought rolled, like oats. It is best used in combination with other rolled or flaked grains as it doesn't soak up as much liquid as oats, and therefore has quite a chewy texture.

Look: Rolled triticale grains resemble large porridge oats and are deep brown in colour.
Taste: A rich taste, stronger than that of oats.
Uses: In the same way as oats, namely in porridge, flapjacks, etc.
Nutrients: Triticale is high in dietary fibre, with high quantities of iron, zinc, folate and magnesium.

Mexican bean and quinoa cakes

I wanted to include an alternative use for South America's wonder grain (or seed, in fact), rather than the usual recipes using quinoa for salads or accompaniments. And what could be better than to combine it with one of Mexico's staple beans to produce these mildly spiced and very healthy veggie burgers – they can be a great way to help children to eat healthily, too, though you might want to reduce the spices slightly and perhaps add some sweetcorn niblets to the mixture as well. You can also serve them in smaller patties as a starter or nibble with a soured cream or salsa dip. They freeze beautifully and can be cooked from frozen.

Bring the stock to the boil, add the quinoa and sweet potato and cook for about 10 minutes, adding the kidney beans for the final 5 minutes. When the sweet potato is tender and the quinoa cooked, drain thoroughly and cool.

Put all the remaining ingredients (except the olive oil) in a bowl. Add the cooled bean and quinoa mixture and, using the end of a rolling pin or a masher (I prefer the rolling pin), squash the mixture, crushing the beans and potato as you go. Season with salt and ground pepper. With wet hands, form into 8 medium-sized or about 15 small cakes. Put onto a plate and chill for at least 40 minutes.

Heat half the oil in a frying pan and gently add half the cakes. Cook over a low-medium heat for about 3 minutes each side, turning them when the underside has browned. Then repeat with the next batch.

Serve with a blob of soured cream, and an avocado and tomato salad alongside.

Serves 4

500ml vegetable or chicken stock
100g quinoa (I like to use the rainbow kind)
300g sweet potato, peeled and cut into 2cm cubes
400g tinned kidney beans, drained and rinsed
1 small carrot, grated
4 spring onions, finely chopped
2 medium free-range eggs
2 small garlic cloves, crushed
½ teaspoon paprika
¼ teaspoon cayenne pepper (less if you don't want any kick)
a handful of coriander, chopped
juice of 1 small lime
2 tablespoons olive oil, for frying

To serve:
soured cream and an avocado and tomato salad
lime wedges (optional)

Congee

Congee or conjee (many names are used) is a sort of rice soup or porridge commonly eaten in Asia. It can be eaten in bowls for breakfast and is also served as a main and side dish. In its simplest form, rice (either short- or long-grain, depending on the country) is boiled in a large pot or rice cooker in plenty of water until very soft and oatmeal like. Stock can also be used instead of water and a variety of flavourings and additional ingredients can be added depending on the culture. Poached chicken and fish are popular, as well as sliced spring onion, ginger and sesame oil. Japanese congee or *okayu* is rather more thick than other Asian versions of congee. It is often given to the sick and to weaning children as it is so easy to eat.

Bibimbap

This classic Korean dish consists of white rice surrounded by a colourful array of hot and cold seasoned vegetables, sesame seeds and a spicy red pepper paste. It is commonly topped with an egg and diners often mix it all together before eating. A pile of thinly sliced marinated and seared beef is often served with the bibimbap, too.

Ross il Forn

This is a classic Maltese dish often made in large quantities for celebrations and served both hot and cold. It consists of rice mixed with a sauce made from beef mince, spices and large quantities of tomato paste and eggs. It is then slowly baked in the oven, producing a firm and very much sliceable lasagne-style dish.

Chilli-glazed salmon and ginger parcels

Red rice adds colour and a delicious nuttiness to this recipe. However, brown Basmati is also a suitable alternative. Add some extra chilli sauce or a chopped chilli, if you like a lot of heat, as the dressing is only mildly spiced. Ovens and salmon thickness vary, hence I have kept the cooking window for the fillets quite wide. Check after 15 minutes, but be prepared for them to take a little bit longer.

Take the salmon fillets out of the fridge to bring up slightly in temperature.

Put the rice in a pan with half the ginger and add 500ml boiling water. Cook the rice for the stated cooking time on the packet, minus about 10 minutes, then drain. Meanwhile, preheat the oven to 200°C/180°C fan/gas mark 6.

Put two large squares of foil on the work surface and divide the rice between them. Place two pieces of salmon on top of each and pile the vegetables on top. Mix the dressing in a cup with the remaining ginger and pour over both piles, then seal with another piece of foil on top.

Place on an oven tray in the oven for 15–25 minutes (check after 15 minutes as you don't want them to overcook). Spoon the rice and vegetables onto plates and place the salmon on top. Pour over any juices.

Alternative to red rice: brown Basmati rice

Serves 4

4 salmon fillets (each weighing approximately 150g)
225g Thai red cargo, Wehani or Camargue red rice
2.5cm piece fresh ginger, sliced into juliennes
150g baby sweetcorn, halved lengthways
½ large red pepper, cut into thin strips
100g shiitake mushrooms, thinly sliced
150g sugar snaps, destrung and halved

For the dressing:
2 tablespoons dark soy sauce
2 tablespoons sweet chilli sauce
2 teaspoons sesame oil
2 teaspoons honey
juice of 2 limes
2 garlic cloves, crushed

Oat-coated mackerel with beetroot relish and watercress salad

The Scottish like to use oats as a coating, and mackerel is the perfect oily fish to put with them. I prefer finer oats for this, but you can use the larger cut variety if you prefer. Serve the fish with green beans and new potatoes and a leafy salad with plenty of lemon in the dressing. For a quicker version, just use ready cooked, unpickled beetroot.

To make the relish, put the beetroot into a pan, cover with water and boil for an hour or until tender. Drain and cool, then peel and chop into 1cm cubes and place in a bowl. Add the remaining ingredients, plus some seasoning and stir together. Chill.

For the mackerel, make sure the fillets are washed, then pour the milk into a shallow dish and dip the fillets in so that both sides of the fish are damp. Mix the flour and oats with some salt and pepper and coat the fish fillets on both sides, pressing the fish into the coating so that it sticks (you will find that the skin side doesn't get so much of the oatmeal). Heat the oil and butter together in a frying pan and fry the fillets gently for 5–6 minutes on each side, depending on their thickness.

Meanwhile, toss together the watercress, pumpkin seeds, lemon juice and olive oil with some seasoning. Serve immediately with the mackerel, relish and plenty of lemon to squeeze over.

_____

Alternative coating for fish: cooked quinoa

Serves 2

2 large mackerel fillets
a little milk, for coating
1½ tablespoons plain flour
6 tablespoons fine oatmeal, such as quick-cook porridge oats
2 tablespoons sunflower or vegetable oil
a large knob of butter
sea salt and freshly ground black pepper

For the relish:
1 medium beetroot
1½ tablespoons crème fraîche
½ teaspoon wholegrain mustard
1 heaped teaspoon creamed horseradish
1 tablespoon olive oil

For the salad:
2 handfuls of watercress, larger stems removed
1 teaspoon pumpkin seeds
a squeeze of lemon juice
a drizzle of extra-virgin olive oil

To serve:
lemon wedges

How to make a risotto

Risotto hails from Northern Italy, the key area for growing rice in the country, in particular the high-starch rice that gives risotto its characteristic creaminess. The first recipe for risotto appeared in a cookbook written by Pellegrino Artusi, published in 1891, and the core ingredients and method have not changed. To cook risotto successfully, the first important job is to find the right risotto rice. The three principle choices are Carnaroli rice (generally considered a favourite for a light and creamy risotto), Vialone Nano (good for more robust risottos) and Arborio (some find this the hardest to get right, although generally it is the most commonly found risotto rice outside Italy). The plump grains will suck up all the stock and then release their starch, producing a wonderful silky end result while still retaining a little bite. Be sure to add enough hot liquid so that the finish is creamy, light and smooth, rather than stiff and stodgy. Also, make sure to toast the grains gently with the onions in the oil (a feature risotto shares with both paella and pilaf) but do not let them burn otherwise the rice grains will seal, and not soak up the liquid. Finally, be sure to beat butter and cheese into the creamy risotto (this is called a *risotto mantecato* in Italian) and then leave to rest for a few moments before serving. In my opinion, when it comes to magnificent comfort food, a great risotto should be up there with roast chicken and shepherd's pie!

Basic risotto recipe

All traditional, stove-top risottos should follow the same simple steps. (The quantities are for 4 people.)

1. Fry 1 chopped onion and 2 chopped sticks of celery in a little olive oil over a gentle heat until soft. A little garlic can then be added for the final 1–2 minutes of cooking.
2. Then add 350g risotto rice. Stir the rice over a medium heat for a couple of minutes (this will toast the grains in the oil, but the rice should not brown). When it 'sings' you are ready to add the liquid.
3. Pour in 100ml white wine and cook until evaporated, then gradually add hot stock by the ladleful (about 1.2–1.4 litres will be needed in total, so keep the stock at a gentle simmer), stirring until each ladleful is absorbed before adding the next.
4. When the rice is just softened and there is enough stock to create a soupy rather than stodgy finish, the risotto is ready. Stir in a chunk (about 50g) of butter and about 75g freshly grated Parmesan. Letting the risotto rest for a couple of minutes will make it all the more creamy and delicious.

Flavourings to try in any combination you fancy:

Asparagus – add chopped stems at stage 1 and blanched tips in the final 5 minutes of stage 3.
Bacon – add chopped bacon and fry at stage 1.
Chicken – fry chicken breasts until just cooked through and slice before piling on top of the risotto.
Courgettes – fry in a little olive oil and add to the risotto at stage 4.
Crabmeat – add cooked white crabmeat for the final 5 minutes of stage 3.
Herbs – add the chopped leaves of woody herbs at stage 1 or soft herbs at stage 4.
Mushrooms – fry in a little olive oil and add to the risotto in the final 5 minutes of stage 3.
Sausage/chorizo – slice and fry at stage 1.
Spinach – stir a mound of washed spinach leaves into the risotto at the end of stage 3.

Prawn, pea, lemon and mint risotto

A wonderfully light risotto with fresh summery flavours, which needs only a green salad as an accompaniment. To avoid it being too heavy, make sure that you add enough stock to make it quite runny in texture. Adding the butter at the end and resting is crucial as it allows the rice to relax.

Heat the olive oil and butter in a pan. Add the onion, celery and garlic and gently soften for about 10 minutes. Then add the rice, stir over the heat to start to toast the grains (a minute or so), then pour in the wine and cook until the alcohol evaporates (again about a minute).

Heat the stock in a pan and keep warm. Add about 300ml of the stock and some seasoning to the rice. Stirring occasionally, simmer over a medium heat until the liquid has nearly all gone. Add another 300ml of hot stock and repeat the process, adding more stock as needed until you are down to the final 200ml of stock.

Taste the rice and if it is not very nearly cooked add a little more stock and cook for a further few minutes – it should be fairly liquid rather than stiff. Add the peas and prawns and cook for another 3 minutes until the prawns are pink. Then throw in the Parmesan, butter, lemon zest and juice. Stir, turn off the heat, then leave to rest for 5 minutes before stirring in most of the mint. Serve, garnishing with the remaining mint.

Serves 4 generously

1½ tablespoons olive oil

25g butter, plus a large knob of butter for the end

1 onion, finely chopped

2 sticks of celery, finely chopped

1 large garlic clove, chopped

350g risotto rice, such as Carnaroli, Vialone Nano or Arborio

100ml white wine

1.3–1.5 litres chicken or vegetable stock

250g frozen peas, defrosted

20 raw, peeled tiger prawns

3 tablespoons freshly grated Parmesan

zest and a good squeeze of ½ lemon

2 tablespoons chopped mint

sea salt and freshly ground black pepper

Seafood paella

This classic Spanish dish of rice with seafood or, in some cases, rabbit, duck or chicken, originated in Valencia and takes its name from the large, round shallow-sided pan in which it is cooked. In ancient traditions, the pan would become the table around which guests would sit, eating directly from the pot. This recipe comes from Sam and Eddie Hart, owners of Fino and Quo Vardis in London. They are experienced paella tasters and cooks and their paella knocks the socks off any others I've tried.

To make the stock, heat the olive oil over a low-medium heat in a saucepan large enough to hold 4 litres and cook the prawn shells and heads for 1–2 minutes. Break the shells up with a wooden spoon. Add the onions and fennel and sweat over a medium heat for 10–15 minutes until starting to caramelise.

Add the tomatoes, tomato purée and paprika and cook for 5 minutes. Then add the brandy and light immediately, burning off all the alcohol. Deglaze the pan, lifting all the caramelised ingredients from the bottom with a wooden spoon.

Add the thyme, bay leaves and 4 litres of water and bring to the boil. Reduce to a simmer and reduce the liquid until just over half remains (this will take 40 minutes to 1 hour). Strain the liquid through a sieve and discard all the solids. Season with the salt and some freshly ground pepper and keep to one side if using straight away, or chill in the fridge until ready to use.

For the paella, heat the olive oil in a large paella pan (with a width of about 15cm) over a low-medium heat and fry the onions for 2 minutes. Add the peppers and salt and fry for 2 minutes, then add the garlic and fry for a further 2 minutes. Add the bay leaves and squid and cook for 4 minutes, then the paprika and cook for 2 minutes, stirring well. Add the rice, mix well and cook for 2 minutes. Pour in 1 litre of the boiling hot stock, sprinkle in the saffron and bring back to the boil over a high heat. Simmer for 10 minutes.

Add the prawns, placing them in a ring towards the middle of the pan. Add the mussels (pushing them down into the rice) around the outer edge of the pan then put the clams (pushing them down into the rice) in between the prawns and mussels. Cook for a further 10–15 minutes, shaking and moving the pan over the flame and pushing the seafood under the surface every so often to ensure even cooking. If it looks as though the paella is drying out, ladle a little more stock over the rice (always remembering that you are aiming for a dry not a soupy result). Taste for seasoning. Remove from the heat and cover with foil and rest for 5 minutes. Remove the foil, sprinkle with parsley and a drizzle of extra-virgin olive oil. Serve immediately with lemon wedges on the side.

Serves 8–10

For the stock:
4 tablespoons light olive oil
shells and heads of the tiger prawns (see below)
2 small onions, chopped
2 small fennel bulbs, trimmed and chopped
3 small tomatoes, chopped
3 tablespoons tomato purée
2 teaspoons sweet paprika
150ml brandy
2 sprigs of thyme
4 bay leaves
2 tablespoons Maldon salt

For the paella:
5 tablespoons light olive oil
2 large onions, finely chopped
2 red peppers, finely chopped
2 green peppers, finely chopped
2 pinches of Maldon salt
2 garlic cloves, sliced lengthways
4 bay leaves
500g cleaned, fresh squid, cut into rings
2 teaspoons sweet paprika
800g Bomba paella rice
2–2.25 litres prawn stock (see above)
4 small pinches of saffron
1kg (approximately 16) medium tiger prawns, heads and shells removed and kept for stock
600g fresh mussels
600g fresh clams
1 small bunch of flat-leaf parsley, finely chopped
2½ tablespoons olive oil
freshly ground black pepper
4 lemons, cut into wedges

One-pot Sicilian fish stew with couscous

This delicious, summery, one-pot stew is based on the classic recipe *couscous alla trapanese*, hailing from Trapani in Sicily. The island of Sicily lies close to Tunisia and the recipe intertwines the traditions and foods of both cultures, combining couscous, the staple food of North Africa (usually served with meat and vegetables in tagines) with the wonderful fish and seafood and Mediterranean flavours that are common to Sicilian cooking. The traditional recipe takes many hours to prepare. This is a simplified version but still uses the key ingredients – fish, fresh tomato sauce and couscous. If you prefer, you can cook the couscous separately and then stir in a ladleful of the tomato sauce as well as some chopped parsley – but personally I don't see the point of adding to the washing up.

Heat the oil in a large saucepan and add the onion and celery. Fry gently for 5 minutes, then add the pepper, chilli and garlic and cook for another 5 minutes. Add the anchovies and white wine. Cook until the wine has evaporated, then add the tomato purée and stir over the heat for 3 minutes. Pour in the boiling hot stock and tomatoes, add two pinches of sugar and a little salt and pepper, and bring up to the boil. Simmer for 15 minutes, covered with a lid.

Meanwhile, just cover the couscous with 100ml boiling water and cover with clingfilm. Once the tomato sauce has cooked for its time, stir in the couscous, fish and seafood, and cook for a further 5 minutes. Sprinkle with the parsley and lemon zest, and serve.

Serves 4

1½ tablespoons olive oil
½ large onion, chopped
2 sticks of tender young celery, trimmed and sliced
1 red pepper, cut into chunks
1 large mild red chilli, deseeded and finely chopped
3 large garlic cloves, chopped
4 anchovy fillets
a splash of white wine
2 teaspoons tomato purée
400ml fresh fish or chicken stock
1 x 400g tin chopped tomatoes
a little sugar
100g couscous
550g red snapper fillet, skin removed and cut into chunks
250g raw, peeled king prawns
350g mussels, cleaned
2 tablespoons chopped flat-leaf parsley
zest of ½ lemon
sea salt and freshly ground black pepper

Salmon, dill and caper fishcakes

Adapt this as you like – you can use smoked haddock instead of the salmon, peas or sweetcorn instead of the beans, or different herbs to suit your taste or your fancy. Semolina is a great storecupboard alternative to breadcrumbs – it requires no preparation and still produces a crisp coating.

Using a large steamer, bring the water in the base to the boil with a little salt and add the potato chunks. Place the fish, skin-side down, into the steamer section and season with salt and pepper. Cook both for 10–12 minutes or until the fish is just cooked, adding the green beans to the steamer for the final 4 minutes until they are tender but still have some bite.

Remove the fish and beans layer and set to one side to cool off a bit, before chopping the beans into 1 cm pieces. Drain the potatoes thoroughly, transfer to a large bowl and add the butter and a little more salt and pepper. Mash together until smooth. Stir in the egg yolk, the tablespoonful of flour, chopped beans, capers, herbs, spring onion, cayenne pepper and lemon juice. Then flake in the fish and gently fork together until combined. Taste, adding seasoning or lemon as needed. Chill in the fridge for an hour.

Sprinkle some flour onto one plate, beat the remaining egg in a bowl and sprinkle the semolina onto another plate. Form the fish mixture into 6–8 cakes, depending on how large you want them to be. Coat in the semolina, dip in the beaten egg and lastly coat with the semolina. Leave in the fridge until ready to use (up to 3 hours ahead is fine).

Heat the oil in a frying pan and gently fry the cakes until golden on both sides – about 3–4 minutes per side. Serve with the herby mayonnaise and dressed salad leaves.

Alternative to semolina coating: cooked quinoa or couscous, or breadcrumbs

Serves 4

300g potatoes, peeled and
   cut into 5cm chunks
350g salmon, smoked
   haddock or cod fillet, or a
   mix of the two, skin on
100g green beans, the
   thicker not the fine type
a large knob of butter
1 whole egg plus 1 egg yolk
1 tablespoon plain flour, plus
   a little for dusting
1½ tablespoons capers,
   rinsed and chopped
1 tablespoon chopped
   dill or flat-leaf parsley
3 spring onions, very
   finely chopped
a pinch or two of cayenne
   pepper
a good squeeze of lemon
   juice
3–4 tablespoons semolina
vegetable or mild olive oil,
   for frying
sea salt and freshly ground
   black pepper

To serve:
mayonnaise mixed with a
   squeeze of lemon and some
   chopped fresh herbs
salad leaves

Wheat
Latin name: triticum

Wheat, the world's oldest domesticated grain, is more widely grown than any other cereal – though it actually ranks second to rice in terms of its importance as a human food. The two most widely grown varieties are common wheat, *Triticum aestivum*, which accounts for 95 percent of all the wheat consumed in the world today, and durum wheat, *Triticum turgidum* ssp. *durum*, used to make pasta. The most important characteristic to have influenced wheat's hegemony in the cereal world is the fact that it contains gluten, the protein that gives wheat grains their unique breadmaking quality.

Einkorn (*triticum monococcum*) was almost certainly the first type of wheat to be domesticated, in the so-called Fertile Crescent in around 12000BC. Over time, early farmers adopted the grain as their staple food and wheat cultivation spread from the Middle East and became established across Asia and Europe. Nowadays, China, the United States, India, Russia and Canada are the five main producers of wheat, which grows best in temperate climates. It can withstand fairly harsh environments, however, and can be grown in areas that are too dry or too cold for the more warm weather-affected grains such as rice and corn. As well as being nutritious, wheat is easily stored and transported, and processed into various types of food, making it one of the most useful and versatile grains.

Nowadays, wheat is classified according to whether the crop is hard or soft, planted in winter or spring, and red or white in colour. These variables impact how the wheat can be used as well as the yield. Hard wheats have the highest levels of gluten, high enough to make bread dough rise during fermentation and, in the case of durum wheat, high enough to give pasta its firm shape. Soft wheats contain less protein and are better for making tender pastry and biscuits.

Emmer wheat (farro) and spelt
Latin name: Emmer wheat/farro – triticum dicoccum; Spelt – triticum spelta

These members of the wheat family are very similar and, along with einkorn, are the ancient precursors of our modern grain. Emmer wheat is also known by its Italian name, farro, though confusingly Italians also use farro to describe spelt and other ancient wheat grains; as a general rule, 'farro medio' refers to emmer wheat, and 'farro grande' to spelt.

Emmer wheat emerged from Egypt in around 10000BC and is considered the world's second domesticated grain after einkorn. Spelt, an evolutionary cross between wild emmer wheat and goat grass, emerged several millennia later, also in the Fertile Crescent. Both forms of wheat spread around the world from there, including northwards into Europe. They became particularly popular among the ancient Romans – indeed farro remains a popular ingredient in Italy to this day. Spelt was famously dubbed the 'marching grain' by the Roman Legions on account of its high energy content.

Spelt was an important staple in parts of Europe from the Bronze Age to medieval times but, as mass farming techniques gradually developed, spelt and emmer were increasingly bypassed. They proved lower yielding than more modern wheat varieties and the close-fitting husk around the grains, which traditionally protected the crops from disease, made threshing difficult. Modern wheat, although less resistant to disease, lent itself more readily to mechanisation and became the staple. The benefit of this neglect is that both spelt and emmer wheat remain genetically pure by comparison. They are also finding a growing audience thanks to their nutritional profile and their low gluten content, which means they can sometimes be tolerated by people with wheat sensitivities.

Both grains can be bought in the form of wholegrain or semi-pearled or pearled grains, and used in a similar way to pearl barley. Look out also for products such as couscous, bulgar and cracked wheat made using these alternative wheats – this could be good news for those with intolerances.

Look: Pearl-sized, creamy grains (brown in colour if whole).
Taste and Texture: Distinctive but mellow nutty flavour with a slight sweetness and a firmish nubbly texture.
Uses: When cooked in their whole or pearled form farro or spelt grains remain separate, and al dente, making them perfect in soups, salads, risottos or pilafs. Or use as a stuffing for chicken or vegetables.
Nutrients: Spelt is packed with complex vitamins and iron and has been credited with boosting the immune system, reducing cholesterol and warding off diabetes. It also has a higher protein content than ordinary wheat and contains all nine essential amino acids. Farro is high in fibre and complex carbohydrates, also with a greater protein content than common wheat.

Freekeh
Other names: farika, firik, frik, frieka, freek

Move over quinoa, this young green wheat is the newest and hippest grain on the scene, though it's actually been around for thousands of years and is even mentioned in the Bible (Leviticus 2:14). Freekeh is made from young durum wheat which is smoked or roasted then polished to rid the grains of their shells and cracked to varying levels of coarseness; the resulting texture is similar to that of bulgar wheat. Freekeh has a unique taste, different from all other grains, due to its smoky flavour.

According to legend, in around 300BC, soldiers somewhere in the Eastern Mediterranean region set fire to fields in order to destroy the local people's wheat crop. Trying to save whatever they could, the locals collected the burnt grain. Rubbing off the burnt shells, they found delicious green grains inside which were not only edible but good to eat; they became known as 'freekeh', which means 'the rubbed one' in Aramaic, the ancient Semitic language of the Middle East. The people of the Eastern Mediterranean region, including Egypt, Jordan, Syria and Lebanon, have eaten freekeh ever since. The southern Lebanese region of Jabal Amel used to be famous for the high quality of its freekeh, but it is not easy to find. Today, it is more common to find shops stocking industrially produced roasted freekeh from Syria and Australia.

Look: Pale green/brown nibbly grains of varying coarseness.
Taste and Texture: Smoky, nutty, rich and crunchy.

Uses: Traditionally, in Lebanon, freekeh is added to soups and stews, but it can also be cooked like barley or rice. It is becoming increasingly popular as an ingredient in salads and pilafs, and also works well in stuffed vegetables; lamb neck stuffed with freekeh is popular among the Arabs of Galilee.

Nutrients: Because the grain is picked while still green, it has more minerals and vitamins than if it were picked later, and four times the amount of fibre than brown rice. Freekeh is also much higher in protein than normal wheat and relatively low in carbohydrate. It is rich in calcium, potassium, iron and zinc as well as prebiotics.

Bulgar wheat and cracked wheat
Other names (bulgar): bulghur, burghul or bulgur

Bulgar and cracked wheat look and taste similar, but are created using slightly different processes – the main difference being that bulgar is pre-cooked while cracked wheat is not. Both are most commonly made from durum wheat.

Transforming wheat into bulgar is an ancient process that originated among the Babylonian, Hittite and Hebrew peoples over 4,000 years ago, and may even constitute man's first processed food. It has been a staple of Middle Eastern and Eastern Mediterranean cuisine for thousands of years, including in Turkey where the name 'bulgur' originated. Bulgar wheat is mentioned in the Bible, under the name 'arisah', and the term is still used in parts of the Middle East. Biblical archaeologists have described bulgar as a porridge or gruel prepared from parboiled, sun-dried wheat. This ancient preparation process continues to be used today in some villages: the whole wheat berries are boiled in vast pots until thoroughly cooked, then spread out in the sun to dry. The hardened kernels are then cracked and sieved into different sizes for various uses.

Bulgar can vary not only in how coarsely it has been ground but also in how much of the bran has been removed (which is why some versions look more brown than others). The grains soften quickly either by being soaked in boiling water or stock or just minimal cooking. Unlike bulgar, cracked wheat is not pre-cooked first. It is made simply by milling the wheat berries coarsely so that they crack into smaller pieces. It is then cleaned, husked and processed to the required size. As cracked wheat is not parboiled it takes a little longer to soften than bulgar and needs to be boiled. Once cooked, however, it is pretty much interchangeable with bulgar.

Also, look out for Maftoul (or Palestinian couscous), which is bulgar wheat that has been hand-rolled and sun-dried by women from co-operatives in Palestine.

Look: Nubbly, pale brown grains.
Taste: Mild, slightly nutty flavour.
Uses: Bulgar wheat is most famously used in Middle Eastern tabbouleh salad but it is very versatile and works well in pilafs, salads and also makes a great addition to soups and stews.
Nutrients: Bulgar and cracked wheat are rich in protein, vitamins and minerals as well as high in fibre.

Kamut (khorasan wheat)
Latin name: Triticum turgidum

This ancient variety of hard wheat – and a close relative of emmer wheat – has a kernel twice the size of common wheat and a much richer, nuttier flavour. Unusually, it is produced and sold under the trademarked Kamut® brand, developed by Montana-based farmer Bob Quinn. It is possible that this ancient grain, like many other ancient varieties, originated in the Fertile Crescent of the Middle East; Khorasan was a Persian province that is now part of Iran. But the grain's recent history is perhaps more intriguing than its ancient origins. A handful of khorasan grains first made their way to Montana in the 1940s, the story being that the kernels were taken from an ancient Egyptian tomb – hence the grain's nickname 'King Tut's wheat' and ultimate brand name (which derives from an ancient Egyptian word for wheat). The first attempt to grow and sell khorasan commercially didn't work, but in the 1970s Bob Quinn decided to revive the grain.

Kamut is now a protected species grown exclusively with organic farming methods and under tightly controlled conditions in over 40 countries. It is used in a huge variety of foods available worldwide, including cereals, snacks, breads, pasta, even beer and coffee. For those wishing to cook with it, khorasan is available most commonly as whole grains, but can be also bought as flour, couscous, bulgar wheat, green kamut, etc. and has a similar cooking time.

Look: Long, reddish-brown grains.
Taste: Rich, creamy, toasty flavour which some people describe as rather like that of popcorn!
Uses: Kamut is extremely versatile as it can be bought in so many forms. Try using in pilafs, side dishes and salads or adding to stews and soups.
Nutrients: Kamut is higher in protein than common wheat and is also a good source of vitamins and minerals such as selenium. Though not suitable for coeliacs, research has shown that it is tolerated by many who are intolerant to common wheat.

Semolina
Semolina is the heart (or endosperm) of the wheat kernel that is separated from the grain and cracked into coarse pieces (or middlings) then milled to various degrees of coarseness – the size varying according to the use to which it is going to be put. Semolina can be made from a variety of grains but is most commonly made from durum wheat and used to make pasta; the high protein content of the semolina means that the pasta absorbs less water so cooks without falling apart.

Semolina is thought to have first been produced in the southern Mediterranean basin or perhaps further south in Abyssinia (modern Ethiopia). It was produced in Egypt during the Byzantine period (4th to 7th centuries AD) and became popular in North Africa as the basis of dishes such as couscous. The word semolina derives from the Latin word *simila* meaning 'flour', though it may have had Sanskrit origins before that. Wheat semolina is, in fact, very popular in India and sweet semolina puddings cooked with other 'pure' foods such as milk, sugar, ghee and bananas are served as a divine gift (called Prasad) during religious ceremonies.

Look: Pale yellow in colour, resembles sand.

Taste and Texture: Smooth and creamy, with a mild flavour.

Uses: Semolina is wonderfully versatile, being great in milk-based puddings, cakes, pasta, pizza, bread and biscuits. You could also try it as a delicious polenta-style accompaniment, made with milk and mixed with Parmesan, or as a coating for fish cakes or goujons, or for thickening soups and stews. In India it is also made into a batter and fried on a griddle to accompany curries.

Nutrients: Semolina is low in fat, high in carbohydrate and contains plenty of protein and several important minerals as well as vitamins E and B. It is low in sodium and fairly high in fibre.

Wheatgerm

The hero of all health food stores, wheatgerm is one of the most nutritional products available. It contains 23 nutrients in total, offering more nutrients per ounce than any other vegetable or grain. As its name suggests, the wheat germ is the part that germinates to form a new plant; although constituting only around 3 percent of the grain, the germ contains most of the wheat's nutrients – except, of course, the fibre. As well as being eaten as an ingredient in wholemeal bread, wheatgerm is sold separately as a health food, and is also valued as a dietary supplement (in oil form) and as an ingredient for skincare products owing to its abundance of B vitamins.

Manufacturers discovered how to separate the germ from the kernel in the 19th century, which was a major step in the wheat refining process as the oil-rich germ reduced the shelf life of flour. The relatively short life of wheatgerm makes it unusual among grains. You can buy wheatgerm separately either as raw or toasted flakes. The raw flakes should be kept in a sealed container in the fridge. The toasted flakes have a longer shelf life but also have a slightly lower nutritional content.

Look: Tiny flakes, mid-brown in colour.

Taste: Mild and easy to incorporate in recipes. The toasted flakes are rather more nutty in flavour.

Uses: The flakes can be added in small amounts to add a distinctively nutty flavour to baked goods or as a healthy topping. Scatter them over yogurt, fruit or cereal, add to porridge or smoothies, or add to bread or cakes.

Nutrients: wheatgerm is high in just about everything, including fibre, and is very high in protein, potassium and iron, and also contains riboflavin, calcium, zinc, magnesium and vitamins A, B (especially folic acid) and E.

Wheat berries

These are not really berries, but the whole wheat kernels from which bread and flour are made. They contain all three parts of the grain, including the germ, bran and starchy endosperm; only the husk, the inedible outer layer, has been removed. Wheat berries are usually named after their growing season (winter or spring), gluten content (Hard or Soft) and colour (Red or White) – but the good news is that they are pretty much interchangeable when it comes to cooking and using them in recipes. The berries are often sold in pre-cooked form, under the French brand name Ebly, which reduces the cooking time to just 10 minutes and dispenses with the need for soaking. Wheat berries are popular in France, as a side dish, and they are also the main ingredient in the Russian porridge *kutya*, eaten primarily at Christmas.

Look: Round whole beads, similar in size to pearl barley or spelt, the colour varying from light brown to red.
Taste and Texture: Cooked and eaten whole, wheat berries have a wonderfully chewy texture and a subtle nutty flavour.
Uses: Wheat berries are good vehicles for bold salad dressings, yet are delicate enough to go well with milk, honey and cinnamon. The grains hold their shape and chewy texture even after long cooking so are great in soups and stews. Also try with vegetables and dried fruit such as cherries or cranberries.
Nutrients: Like all whole grains, wheat berries get the nod for their exceptional nutrient profile; being unprocessed, they retain the nutrients and roughage that refined wheat products have lost. Thus, they are high in fibre, low in calories and packed with vitamins and minerals. A serving of cooked wheat berries is chock full of manganese, selenium, phosphorus and magnesium. Wheat berries also contain lignans, phytochemicals thought to guard against breast and prostate cancers. Wheat-berry sprouts are bursting with cell-protecting antioxidants, vitamin E and magnesium, which promotes healthy bones and muscles.

For couscous, see page 142.

Thai-fried rice with prawns and pineapple

As popular as the sandwich in the UK, fried rice is often enjoyed by Thais as a lunchtime stop-gap. Sweet honey pineapples grow all around in fields along the roadside and are a welcome addition to this spicy rice dish. The Thais prefer to scoop out the flesh of the pineapple and then serve the fragrant stir-fried rice in the scooped out shell, but I prefer the method below for eating at home. You can also omit the prawns and add more vegetables or some chicken if you prefer.

If you like, make a small slit in the prawns halfway down and push the tails through. Not only do they look neater like that, according to the Thai chef they are also much more tender this way.

Heat half the oil in a wok and add the prawns. Stir-fry over a high heat until pink, then remove and set to one side in a bowl. Add a further tablespoonful of oil, turn the heat down to medium-low and stir-fry the onion for 4–5 minutes, before adding the garlic, ginger and Thai paste. Cook for 2 minutes then add the ham and stir-fry for a further 2 minutes.

Add the cooked rice. Break it up and mix with everything in the wok, then pour in the stock and add the peas, soy sauce and pineapple. Stir-fry for a minute or two then add the cashews, fish sauce, beansprouts, spring onions and half of the coriander. Return the prawns to the pan, toss together and taste (add a little more soy if you like). Serve garnished with the remaining coriander and the lime to squeeze over the top.

Alternative to Jasmine rice: cooked and cooled Basmati or wholegrain long-grain rice

Serves 4 generously

700g Thai Jasmine rice, cooked and cooled
20 raw tiger prawns, peeled with tails left on
2 tablespoons groundnut oil
1 red onion, sliced
2 garlic cloves, crushed
1 heaped teaspoon grated fresh ginger
1 teaspoon Thai red paste, or to taste
100g thickly sliced ham, chopped
300ml chicken stock
100g frozen peas, defrosted
1–1½ tablespoons dark soy sauce
½ small pineapple, peeled and roughly chopped
2 heaped tablespoons cashew nuts, roasted and roughly chopped
1½ tablespoons fish sauce
2 handfuls of beansprouts
4 spring onions, trimmed and chopped fincly
a handful of coriander, chopped

To serve:
lime wedges

Jambalaya

I grew up on jambalaya. My mother used to make her own version of the New Orleans dish – probably while dreaming of being a jazz groupie in New Orleans one summer. Jambalaya is a great dish for large numbers as it's dead easy to make and needs nothing but a salad on the side, if that. Vary what you add – leftover chicken or turkey, smoked or garlic sausage or leftover bangers. My favourite version uses chorizo and thick chunks of good-quality cooked ham with plenty of flavoursome prawns and a hint of spice. It's a great Boxing Day recipe for using up the Christmas gammon and turkey: just heat through the poultry for the final 2–3 minutes of cooking, once the rice is nearly cooked. Jambalaya actually benefits from being given a few minutes of rest before serving. You can also make it the day before, slightly undercooking the rice, and just reheat with extra tomato juice or chicken stock. Thanks Mum!

First, heat half the oil in a large heavy-bottomed saucepan or sauté pan and brown the chunks of chorizo or sausage for 3–4 minutes over a high heat. Add the ham and fry for a further 2 minutes. Transfer to a plate.

Cut the onion into quarters. Keep one quarter with the end intact so that it doesn't break up; use this quarter as a pin cushion for the cloves. Chop the remainder of the onion.

Add the remaining oil to the pan and fry the chopped and clove-spiked onion, celery, oregano and thyme for 6 minutes or until soft. Tip in the peppers and stir-fry for 2 minutes before adding the garlic, cayenne pepper and rice. Toss over the heat then add the tomato purée and stir to caramelise. Throw in the bay leaves, pour in the wine and simmer until the wine has evaporated, stirring over the heat.

Add the tinned tomatoes and boiling hot stock, then return the ham and sausage to the pan. Stir, bring to the boil, then simmer, covered, for about 15 minutes or until half the liquid has evaporated and the rice is cooked. Add the prawns and stir for a final couple of minutes. Turn off the heat, fish out the clove-spiked onion and bay leaves and discard. Rest the rice for 5 minutes. Stir in the parsley and serve.

Serves 4–6

- 2 tablespoons vegetable or mild olive oil
- 175g cooking chorizo, smoked pork sausage or spicy sausages, cut into chunks
- 275g thick slices of ham, cut into chunks
- 1 onion, peeled
- 5 whole cloves
- 2 sticks of celery, finely sliced
- ½ heaped teaspoon dried oregano
- 1 teaspoon thyme leaves
- ½ green pepper and ½ red pepper, or just one of either colour, chopped into chunks
- 2 garlic cloves, crushed
- 2–3 pinches of cayenne pepper
- 350g American long-grain rice
- 1½ tablespoons tomato purée
- 2 bay leaves
- 125ml white wine
- 1 x 400g tin chopped tomatoes
- 700ml chicken stock
- 200g cooked, peeled medium prawns, defrosted
- a good handful of chopped flat-leaf parsley

Giant couscous with spicy sausages and roast vegetables

This is a great one-tray autumn supper. Giant couscous – sometimes called Ptitim, Fregola, Israeli couscous or Mograbieh – is pearl-sized balls of semolina which have been toasted, so they need cooking rather than just soaking (as per the smaller granules of traditional couscous). I have suggested wholewheat giant couscous here, but you can use regular giant couscous if you prefer. It will just take slightly less time to cook, so test it after 15–20 minutes.

Preheat the oven to 200°C/180°C fan/gas mark 6.

Put all the vegetables into a baking tray and toss with the spices, oil and some salt and pepper. Bake for 25 minutes, then stir and top with the sausages. After a further 25 minutes (turn the sausages once during this time), stir in the tomatoes, couscous, boiling hot stock and honey. Cook for a further 20–30 minutes or until the couscous is tender. Check the couscous during baking to ensure it doesn't dry out, adding a little extra stock if needed. When it is ready, stir in the parsley and lemon juice and serve.

_____

Alternative to giant couscous: spelt or wheat berries (the latter may need a little longer cooking)

Serves 4

250g parsnips, cut into 4cm cubes

200g sweet potato, cut into 4cm cubes

2 sticks of celery, trimmed and cut into 3cm slices

2 carrots, cut into 4cm cubes

2 small red onions, cut into wedges

3 garlic cloves, kept whole and unpeeled

½ teaspoon cumin seeds

½ teaspoon ground coriander

1½ tablespoons olive oil

8 spicy lamb or pork sausages, merguez or similar

2 tomatoes, roughly chopped

175g wholewheat giant couscous

750ml chicken stock

1 teaspoon honey

2 tablespoons chopped flat-leaf parsley

a good squeeze of lemon juice

sea salt and freshly ground black pepper

Pork fillet with prunes, mustard and barley

This is a delicious variation of the classic combination of pork with prunes and mustard sauce, with the barley and greens cutting the richness. Barley comes in a variety of forms – whole, half-polished, fully polished and so on – and you can use any type for this recipe; just be aware that the semi- or unpolished types will take longer to cook, so extra liquid will be needed.

Heat half the oil in a pan. Fry the onion for 5 minutes, then add the garlic and cook for a further minute before adding the rinsed and drained barley. Stir for a minute, then add the boiling hot stock and bring to the boil. Put a tight-fitting lid on and gently boil for 30–35 minutes until the barley is tender, but with a slight bite. Remove the lid for the final 3–5 minutes to let any remaining liquid evaporate. Meanwhile, soak the prunes in the wine for 30 minutes.

Bash the pork slices to flatten them slightly, then season with a little pepper and salt. Heat the remaining oil in a large frying pan and, over a high heat, brown the pork on both sides. Remove to a dish. Add the kale or greens and fry, adding a splash of water to steam-fry them. After a couple of minutes add the prunes and wine. Reduce for 30 seconds or so, then stir in the mustard and crème fraîche and add the pork back to the pan. Taste for seasoning, then spoon the pork and prunes onto plates. Add the cooked barley to the sauce and stir together. Serve with the pork.

Alternative to barley: pearled spelt or farro

Serves 2–3

2 tablespoons olive oil
1 small onion, chopped
1 garlic clove, crushed
100g barley, rinsed
650ml chicken or vegetable stock
15 ready-to-eat pitted prunes
100ml white wine
400g pork fillet, cut into 2cm thick slices
200g kale or greens, thickly shredded
2 teaspoons Dijon mustard
3 tablespoons crème fraîche
sea salt and freshly ground black pepper

Chinese-style fried rice with chicken

There are endless variations to this Chinese favourite. Most tend to contain cooked white rice, egg and peas. Then the alternatives are endless: try adding chopped roast pork, ham, prawns, peppers, spring onions or beansprouts. I use a little black bean sauce for added flavour, but you can simply season with a splash of soy sauce and sesame oil if you prefer. You can also use cubes of raw boneless chicken thigh if you don't have any leftover roast – add just before the mushrooms, seal for 3–4 minutes, then add the mushrooms and continue to cook as per the recipe.

Heat a wok and add half a tablespoonful of the vegetable oil. Beat the eggs with the sesame oil and soy sauce. Add to the pan and swirl the wok to make an omelette then, when set, turn over and cook the other side. Remove to a plate.

Add a splash more oil to the pan and stir-fry the onion for 2–3 minutes. Add the garlic, chilli, ginger, sliced spring onions and mushrooms and stir-fry for a further 2–3 minutes before adding the peas, chicken, rice and a dash of water. Toss together for 2–3 minutes, then stir in the black bean sauce followed by the omelette, which you can roughly chop before adding, plus some seasoning.

Taste, adding a little more soy sauce if needed, then sprinkle over the chopped green onion stalks and serve.

Variations:
Ham and prawns – stir in three thick slices of ham, chopped, and a handful of cooked small prawns (defrosted) instead of the chicken.

Vegetable – omit the chicken and add a chopped red pepper to the wok with the onions. Stir in two handfuls of beansprouts just before serving.

Roast pork – ideally use Chinese roast pork or some leftover Sunday roast. Chop into small chunks and add instead of the chicken.

Alternative to cooked white rice: cooked wholegrain rice or red rice

Serves 3–4

1–2 tablespoons vegetable oil
2 free-range eggs
2 teaspoons sesame oil
2 teaspoons dark soy sauce, plus extra for seasoning, if required
½ onion, finely chopped
2 garlic cloves, crushed
1 red chilli, deseeded and chopped
5cm piece fresh ginger, grated
4 spring onions, white part sliced and green ends finely chopped
75g shiitake mushrooms
100g frozen peas, defrosted
300g leftover roast chicken
400g cooked white rice, cooled
1 tablespoon black bean sauce, preferably Lee Kum Kee
sea salt and freshly ground black pepper

Baked chicken, bacon and pumpkin risotto

Although baking a risotto is one of those practises that's quite probably frowned upon in Italy, for those of us, like me, who are short on time, it's a great one to have up the sleeve. With a salad on the side this is enough to feed four people, but if you are a greedy family then allow three helpings!

Preheat the oven to 200°C/180°C fan/gas mark 6.

Heat the oil in a flameproof casserole. Add the onion, celery and bacon and fry for 3 minutes. Add the chicken and stir-fry for a further 2 minutes, before adding the garlic and herbs. Add the rice and stir for 30 seconds or so.

Add the wine and cook until it no longer smells of alcohol – about a minute. Then pour in the boiling hot stock, season with a good grind of pepper and bring up to the boil. Stir, cover with a lid and place in the oven. Bake for 25–30 minutes, stirring in the pumpkin after the first 10 minutes. Test to see if the rice and pumpkin are tender, adding a little extra water if the risotto needs an extra 5 minutes in the oven. Stir in the butter and Parmesan. Leave for 5 minutes with the lid off, taste for seasoning and serve.

Variations:
Spanish style – Fry 100g chopped chorizo, ½ teaspoon paprika and 1 sliced red pepper in place of the bacon. Omit the pumpkin, adding 200g cherry tomatoes for the final 10 minutes of cooking.

Smoked salmon (great as a starter for 6) – Omit the bacon, chicken and pumpkin. Stir in 1 scant tablespoonful of lemon thyme or common thyme leaves with the rice and then 200g smoked salmon ribbons when the risotto comes out of the oven. Garnish with a further 100g smoked salmon ribbons and some chopped parsley or chives and lemon wedges.

Healthier baked risotto – Although it goes against everything I say about a traditional risotto, using two types of rice does make this a healthier option. Use half Arborio and half brown Basmati rice . Reduce the stock by 100ml, omit the butter and stir in 100g cooked peas at the end. And increase cooking time to 35 minutes.

Serves 3–4

1½ tablespoons olive oil
1 medium onion, chopped
1 large or 2 small sticks of celery, chopped
2–4 slices of smoked back/ streaky bacon, chopped
450g chicken thighs, skinless and boneless, cut into cubes
2 large garlic cloves, crushed
1 teaspoon finely chopped rosemary or thyme leaves
250g risotto rice, such as Carnaroli, Vialone Nano or Arborio
a good splash (approximately 4 tablespoons) of white wine
850ml well-flavoured chicken stock
500g pumpkin or butternut squash, peeled and cut into 4cm cubes (prepared weight about 400g)
a knob of butter
1½ tablespoons freshly grated Parmesan
sea salt and freshly ground black pepper

Nasi goreng

One of Indonesia's national dishes, Nasi Goreng actually hails from China, but was introduced when the Chinese traded with Indonesia from about 2000 BC. The dish varies from place to place, sometimes using prawns and chicken, sometimes using just vegetables, but is unified by the fact that it always features fried rice and is served with a fried egg on top. This version was taught to me at an Indonesian hawker stall – it's super quick to make and is made all in one wok. I love a handful or two of beansprouts added at the end, but I'm not sure that's done in Indonesia! You can use wholegrain rice for added nutrients. It's best to have all your ingredients prepped and ready to add to the wok for speedy stir-frying.

First, heat half the oil in a wok and fry the eggs until nearly cooked. Transfer to a plate. Keep warm in a low oven.

Add the onion to the wok with a little extra oil if needed and fry until beginning to soften. Throw in the chicken, turn the heat to high and brown all over. Reduce the heat to medium, add the garlic and shredded greens and stir-fry for 1–2 minutes or until cooked but still crunchy. Add the rice, oyster sauce, soy sauce and sesame oil. Season and stir together to heat through. Serve topped with the fried eggs.

Alternative to cooked white long-grain rice: cooked brown Basmati rice

Serves 4

2 tablespoons vegetable oil
4 free-range eggs
1 onion, sliced
400g skinless and boneless chicken thighs, cut into small chunks
2 garlic cloves, chopped
200g greens, such as kai lan, spring greens or cabbage, shredded
400g cooked white long-grain rice, cooled
2 tablespoons oyster sauce
1½ tablespoons dark soy sauce
1 tablespoon sesame oil

Hainanese-style poached sesame and ginger chicken rice with greens

Succulent chicken in a simple but tasty broth, served with perfectly cooked rice, greens and chilli sauce. Though it gets its name from its origins in Hainan, a tropical island off China's southern coast, it was really when emigrant Chinese brought the dish to Singapore that it gained a new personality and became famous. It is seen at every hawker stall in Singapore, and is my 'go to' if I'm wanting to buy or cook something healthy and tasty. I have used chicken breasts here, but you can use a whole chicken if you are cooking for more people.

Put 1.8 litres cold water into a large saucepan. Add the sliced ginger, chilli, spring onions, peppercorns and ½ teaspoon salt and bring to the boil. Season the chicken with salt and add to the boiling broth. Turn the heat to low and simmer for 20 minutes, then turn off the heat. Remove the chicken from the pot and place on a board to rest (traditionally the chicken would be put in an ice bath, but I prefer it served warm). Reserve 400ml of the broth for cooking the rice.

In a small bowl, mix the soy, sesame oil and toasted sesame seeds together for the sesame sauce. Put all the chilli sauce ingredients into another small bowl and mix together. Set both aside.

To cook the rice, heat the groundnut oil and sesame oil in a saucepan over medium heat. Add the ginger, chopped onion and garlic and stir-fry for 2–3 minutes. Be careful not to let it burn. Add in the drained rice and stir over the heat to coat. Add the reserved poaching broth (reheat it first if it has cooled) and a little salt and bring to a boil. Immediately turn the heat down to low, cover the pot and cook for 10 minutes. Remove from heat and let it sit (with lid still on) for 5–10 minutes more.

To serve, heat up the broth, add the greens and cook for 3 minutes or until the vegetables are tender. Remove the skin from the chicken and slice thickly. Serve the chicken, greens and rice in shallow bowls with a ladleful of the hot broth, drizzled with the soy-sesame sauce and garnished with coriander. Pass around the chilli sauce.

Serves 4

6cm piece fresh ginger, sliced
1 red chilli, slit lengthways but kept in one piece
4 spring onions, trimmed and cut into 3cm slices
8 black peppercorns
4 large chicken breasts, skin on
4 heads kai lan, bok choi or pak choi

For the sesame sauce:
2 tablespoons dark soy sauce
2 teaspoons sesame oil
2 teaspoons toasted sesame seeds

For the chilli sauce:
2 tablespoons lime juice
2 teaspoons caster sugar
3 tablespoons chilli sauce
1 fat garlic clove, crushed
5cm piece fresh ginger, grated
a generous pinch of salt

For the rice:
1 tablespoon groundnut oil
1 teaspoon sesame oil
2cm piece fresh ginger, grated
1 small onion, chopped
2 garlic cloves, crushed
250g Jasmine (Thai fragrant) rice, washed and soaked in cold water for 10 minutes
400ml reserved chicken poaching broth
a handful of coriander leaves, chopped, to serve

Roast chicken with sweetcorn stuffing

The wonder of sweetcorn in a stuffing is that it provides little pockets of sweet juiciness, lightening the mixture. Add some tasty gravy and roast spuds and you can't beat it! If you are serving with potatoes or veg, save the cooking water to use in the chicken stock for the gravy – it add nutrients and flavour.

Take the chickens from the fridge 30 minutes before stuffing to bring up to room temperature. Preheat the oven to 200°C/180°C fan/gas mark 6.

Make the stuffing. In a small saucepan, heat the oil and fry the onion gently until soft, adding the garlic for the final minute. Leave to cool, then combine with the remaining stuffing ingredients in a bowl, along with a good grinding of pepper and salt. Mix together until well combined and stuff into the cavities of the chickens.

Season the birds, place into a roasting tin and bake for 15 minutes. Then pour over the wine, turn the heat down to 170°C/150°C/gas mark 3 and roast for a further 1–1¼ hours or until the juices run clear when the legs are pierced with a skewer. Rest the birds for 10 minutes on a board before serving.

For the gravy, pour most of the chicken juices from the roasting tin into the hot stock. Stir the flour into the roasting tin, over a gentle heat and let it all brown a little. Gradually stir in the stock and juices and bring to the boil. Turn down to a simmer and cook for 5 minutes, then taste, adding extra seasoning if needed and serve with the chicken and stuffing.

Serves 6–8

For the chicken:
2 small chickens
100ml white wine
1 heaped tablespoon plain flour, mixed with a little water until smooth
250ml chicken stock

For the stuffing:
2 teaspoons sunflower or vegetable oil
½ small onion, finely chopped
1 garlic clove, crushed
230g sausage meat or pork mince
50g multigrain bread, made into crumbs
zest of 1 lemon, juice of ½
½ x 400g tin sweetcorn, drained
1 medium free-range egg
3 heaped tablespoons chopped flat-leaf parsley
2 sprigs of thyme, leaves removed and chopped
sea salt and freshly ground black pepper

Roast duck legs with Asian-spiced red rice and plum sauce

Red rice is grown in Asia as well as in the Camargue area of France, and its colour and nutty flavour make it the perfect accompaniment to rich and crispy duck legs and tart plums. Taste the plums, and add more sugar to the sauce if it's too tart. Marinating the duck and cooking the rice can all be done ahead of time, to avoid last-minute rushing about!

First, make the marinade for the duck. In a pestle and mortar, pound the Sichuan peppercorns, star anise and fennel seeds until ground, then add the garlic and ginger and continue to grind until a rough paste is formed. Stir in the soy sauce, chilli flakes, sesame oil and vinegar. Put the duck legs into a dish and smother over the marinade. Leave for at least 4–6 hours or preferably overnight.

Soak the rice for 30 minutes in cold water, then discard the soaking water, transfer to a pan and add the hot chicken stock. Bring to the boil and simmer for 25–30 minutes or until just tender (the time varies with different types of rice), topping up with a little extra boiling water if needed. Drain, then cool and keep in the fridge until ready to use.

When you're ready to cook, preheat the oven to 160°C/140°C fan/gas mark 3. Place the sliced onion and plums in the base of a roasting tin, sprinkle with the sugar, add the cinnamon halves and mix together using your hands. Put the duck legs on top with any remaining marinade. Cover with foil and bake for 1½ hours, removing the foil after 30 minutes to allow the duck to brown.

Remove the duck from the oven and increase the temperature to 220°C/200°C fan/gas mark 7. Transfer the duck legs to a baking sheet and put back into the oven for 15–20 minutes to crisp (watch out for excess fat spilling when you do this). While the duck crisps, transfer the plums and all the juices to a saucepan. Leave to settle for 5 minutes, then carefully spoon any fat from the surface. Bubble the plum sauce vigorously on top of the stove for about 10 minutes or until reduced by about a third to a half. Taste and adjust the seasoning, adding extra salt, pepper or sugar if necessary, then keep warm.

Heat a wok on the hob and add the sesame oil. Add the spring onions, garlic and chopped greens, toss together, then add a splash of water and stir-fry for 3–4 minutes or until tender. Add the cold cooked rice and the soy sauce and toss over the heat for 3–4 minutes or until piping hot. Taste, seasoning with salt, pepper and a little more soy if needed, then spoon onto plates. Top with the duck legs and a spoonful or two of the plum sauce, discarding the cinnamon stick. Serve the remaining plum sauce in a bowl to hand around.

Serves 4

4 duck legs, trimmed of some of the excess fat
½ onion, sliced
450g ripe dark red plums, halved and stones removed
1 heaped tablespoon dark brown soft sugar
1 cinnamon stick, broken into two
sea salt and freshly ground black pepper

For the marinade:
½ teaspoon Sichuan peppercorns (use black peppercorns if you can't find any)
1 star anise
½ teaspoon fennel seeds
1 large garlic clove, crushed
3cm piece fresh ginger, grated
2 tablespoons dark soy sauce
2 pinches of chilli flakes
1 teaspoon sesame oil
2 teaspoons rice vinegar

For the rice:
200g Thai red cargo rice or Camargue red rice
500ml chicken stock
1 teaspoon sesame oil
3 bunches of spring onions, trimmed and sliced
1 large garlic clove, sliced
250g greens such as kai lan, pak choi or young spring greens, trimmed and cut into smaller lengths
2 teaspoons dark soy sauce

Slow-roast lamb and bulgar wheat pilaf with saffron yogurt

A great friend, Sarah Randell, held a weekend party for her birthday at Lartigolle in France. On the final day, we ate the most incredible Middle Eastern-style lamb, which had been slow-roasted with home-made pomegranate molasses. Bulgar wheat is the perfect accompaniment, as it is commonly used in pilaf-style recipes, and is an ideal match with the aubergine and herbs to counteract the richness of the meat. Bulgar is also full of fibre and is low-GI, making it the perfect healthy accompaniment to the rich lamb. Eat the leftover lamb and yogurt sauce in flatbreads – delicious! You can use the same marinade for chops, and serve with the same pilaf.

Preheat the oven to 220°C/200°C fan/gas mark 8.

First, make a start on the pilaf. Put the aubergine chunks and the garlic cloves into a roasting tin, drizzle with 1 tablespoon of olive oil and season with salt and pepper. Roast in the oven for 20–30 minutes or until softened and beginning to brown. Remove, cool and set to one side. Leave the oven on for the lamb.

Meanwhile, place the lamb in a roasting dish and pierce all over with the pointy end of a knife. Press the garlic pieces into the holes. In a bowl, mix together the cumin, sumac, molasses, 2 teaspoonfuls of olive oil and a good grinding of salt and pepper. Rub the mixture all over the lamb then place in the oven. After 20 minutes take the lamb out of the oven and reduce the heat to 160°C/140°C fan/gas mark 3.

Pour the wine around the lamb and cover with foil. Bake at the lower temperature for 3½ hours, then remove the foil and finish cooking for a further 20 minutes.

Towards the end of the lamb's cooking time, mix the lemon juice, olive oil and seasoning in a salad bowl. Top with the rocket or watercress leaves and cucumber, but don't toss, just leave in a cool place. Remove the bulgar from the heat, stir in the parsley and taste for seasoning.

Serves 6–8

1 large shoulder of lamb, weighing approx 3.5kg
3 garlic cloves, cut into thick slices
½ heaped teaspoon ground cumin
2 teaspoons sumac
5 tablespoons pomegranate molasses or syrup
2 teaspoons extra-virgin olive oil
200ml glass white wine
seeds from 1 pomegranate
a handful of mint leaves, torn
sea salt and freshly ground black pepper

For the pilaf:
1 large aubergine, cut into 5cm chunks
3 garlic cloves, unpeeled
1 tablespoon plus 2 teaspoons olive oil
a knob of butter
3 red onions, sliced
½ teaspoon cinnamon
½ teaspoon cumin
400g bulgar wheat
675ml weak chicken or vegetable stock
a handful of flat-leaf parsley leaves, chopped

Continue with the pilaf. Put the butter and the 2 remaining teaspoonfuls of oil in a large saucepan and fry the onions slowly for about 10 minutes until soft and tinged with brown. Add the cinnamon, cumin and bulgar wheat and stir together for a minute. Pour in the hot stock (you will need to season with some salt if your stock is home-made), cover and cook gently for 10 minutes or until tender. Add the roasted aubergine for the final 3 minutes.

Once the lamb is cooked, remove to a board and leave to rest loosely covered with foil for 15 minutes. Spoon off any fat on top of the lamb and keep the juices warm in a small pan. Once the lamb has rested, shred onto a warm plate and keep warm in the now cooling oven, covered with foil.

Pour 1 tablespoonful of boiling water over the saffron threads and leave for 5 minutes, then stir into the yogurt with the lemon juice, garlic and oil, plus some salt and pepper.

Spoon the bulgar pilaf onto a warm platter and top with the shredded lamb and any juices. Sprinkle with the mint and pomegranate seeds and serve with the yogurt dressing and the salad, tossed with its dressing.

_____

Alternative to bulgar wheat: cracked wheat, freekeh or long-grain rice

For the rocket salad:
2 tablespoons lemon juice
2 tablespoons extra-virgin olive oil
150g rocket leaves, or watercress broken into smaller sprigs
¾ cucumber, peeled, halved lengthways and sliced
sea salt and freshly ground black pepper

For the yogurt dressing:
2 pinches of saffron
5 tablespoons Greek yogurt
a good squeeze of lemon juice
½ garlic clove, crushed
1 tablespoon oil

Lebanese kibbeh with tahini dressing

The Arabic word *kubbah* means ball, and these crispy morsels of ground meat with bulgar and onion are a classic in Levantine cuisine. The paste is stuffed and then usually shaped into mini torpedoes (a role commonly given to a prospective bride in Lebanon and Syria). It is then fried or grilled and served with salad and a tahini or yogurt dip. Known around the Middle East variously as *kobeiba*, *cubbeh*, *bulgar koftesi* or *kubba*, they can be made bigger or smaller depending on whether you want to serve them as a nibble, a starter or as a main course. I like to eat the kibbeh with crunchy leaves, flatbreads and a Greek-style salad. And I prefer them made with beef but they are most commonly made with lamb or even goat; it's best to use non-lean good-quality mince. You can prepare the kibbeh earlier in the day and keep in the fridge.

Rinse the bulgar wheat three times, then transfer to a bowl and pour over 120ml boiling water. Cover and leave while you prepare the stuffing.

For the stuffing, heat the oil in a pan and fry the onion gently for 8 minutes or until soft. Add the pine nuts and stir over a medium-high heat for a minute or so before adding the garlic, spices, sultanas and lamb and stir-frying until the lamb has browned. Season and turn off the heat, then stir in the coriander leaves.

Drain the bulgar wheat thoroughly and transfer to a bowl. In a processor, pulse the remaining mince, onion and spices together with some more seasoning until combined. Mix into the bulgar, using your hands, as if you are kneading dough. Stir in a tablespoonful of Greek yogurt and season generously.

Take a golf ball-sized amount of the bulgar mix in your wet hands and flatten into your palm. Place a teaspoon or so of the stuffing mixture into the centre and then, with a little more bulgar mix, enclose the stuffing and form into a small lemon shape. Place onto a plate while you make about 12–14 more. Chill for 20 minutes, or longer if you want to make them in advance.

When you are ready to cook the kibbeh, heat 3–4cm or so of oil in a deep frying pan. When sizzling hot but not smoking hot, add half the kibbeh and fry for 7–10 minutes or until they are deep brown all over. Drain on kitchen paper. Keep warm in the oven while you fry the next batch.

Whisk the tahini sauce ingredients together and serve with the kibbeh.

Makes about 15 kibbeh (serving 4–5 people as a main course)

100g bulgar wheat
375g beef or lamb mince
½ onion, roughly chopped
2 good pinches of allspice
½ teaspoon ground cumin
1 tablespoon Greek yogurt
sunflower or peanut oil, for frying

For the stuffing:
1 tablespoon olive oil
1 small onion, finely chopped
2 tablespoons pine nuts
1 garlic clove, crushed
a good pinch of cayenne pepper
2 good pinches of cinnamon
a good pinch of allspice
1 tablespoon sultanas, roughly chopped
125g beef or lamb mince, preferably finely ground
1 heaped tablespoon chopped coriander leaves

For the tahini dressing:
4 heaped teaspoons tahini paste
4 heaped tablespoons Greek yogurt
juice of 1 small lemon
4 tablespoons olive oil
3 tablespoons water
1 fat garlic clove, crushed

French-style lamb and barley stew

Barley has had many uses in many countries, and in parts of Europe including Scotland and Ireland there is a tradition of adding pearl barley to stews, soups and casseroles to make a little go a long way. Aside from the frugal aspect of cooking with barley, I also enjoy the balance it gives, counteracting the richness of the meat and adding extra body to the gravy. If you don't fancy cooking any extra veg, the stew is a great one-pot meal. I like it served with buttered greens. The lamb used in this recipe could be neck fillet, shank middle neck or boneless leg – you will just have to adjust the cooking time to suit the cut. Adding sugar is a tip I found in a number of French recipes for navarin of lamb – it gives the lamb a wonderful colour and slightly caramelises it, too. You can make the stew up to the point of adding the barley and vegetables on day one, then the following day bring it up to a simmer on the stove before adding the barley and vegetables and finishing off in the oven for a further 1–1½ hours.

Preheat the oven to 150°C/140°C fan/gas mark 2. Season the lamb. Heat the oil in a large, oven-proof, heavy-based pan and gently sauté the onion for 8–10 minutes. Transfer to a bowl using a slotted spoon. Then, over a high heat, brown the lamb cubes thoroughly all over, removing to a bowl while you cook the next batch.

Put the onion and lamb back into the pan along with the garlic and sprinkle over the sugar. Toss in the hot fat to colour, add the flour and stir gently over the heat. Throw in the rosemary, pour in a little of the boiling hot stock and scrape all the sediment from the bottom of the pan. Add the tomato purée and cook for a minute. Add the remaining stock and a good grinding of salt and pepper and bring to the boil, then cover and transfer to the oven. After an hour, stir in the pearl barley, chopped tomatoes, carrots and leeks and bring back up to the boil before returning to the oven for a final hour or until the barley and meat are soft.

Ideally let cool, then, using a spoon, skim the fat off the top of the casserole, before gently reheating. Or, if you wish to serve the stew straight away, just leave to settle and try to remove as much of the oil on the surface as possible.

_____

Alternative to pearl barley: wheat berries, pearled spelt or even orzo pasta

Serves 6 generously

1kg boneless shoulder of
  lamb, cubed
2 tablespoons vegetable oil
2 onions, chopped
2 large garlic cloves, peeled
  and squashed with the back
  of a knife
1 teaspoon sugar
1½ tablespoons plain flour
2 large sprigs of rosemary
1.1 litres chicken stock
2 teaspoons tomato purée
100g pearl barley
2 tomatoes, chopped
3 carrots, cut lengthways
  and chopped into chunks
2 leeks, sliced into 4cm
  chunks
sea salt and freshly ground
  black pepper

Desserts, treats & drinks

Stove-top rice pudding – Thai sticky rice with mango – Chocolate and vanilla semolina pudding – Raspberry brûlée rice pots – Black rice pudding with coconut and pandan leaves – Scottish honey and oat praline ice cream – 'Wow your guests' easy soufflés – Cypriot lemon and semolina cake – Dutch apple tart with oatmeal streusel topping – Chocolate-dipped Anzac biscuits – Apricot, chia and coconut flapjacks – White chocolate crispies – Chocolate, cherry and hidden grain fridge bars – Berry shortbread bars – Polenta and ricotta berry torte – Sticky salted caramel popcorn – Lemon barley water

Stove-top rice pudding

While most of us associate rice pudding with a dish baked in the oven, this method of cooking the dish didn't appear until the 17th century. I was introduced to the stove-top, risotto-style method a few years ago, when nutritionist Fiona Hunter made a healthy chocolate version with skimmed milk and cocoa. Even without the whole milk or cream, the starch released from the rice when stirring renders it seriously creamy. Now, I often make rice pudding this way as it is also much quicker. It is a great pudding fix when you are feeling like something sweet, creamy but healthy. Serve with a dollop of jam, a sprinkle of cinnamon and raisins or just as it is.

Put all the ingredients into a pan and bring up to the boil. Then turn down to a gentle simmer and cook, half covered with a lid and stirring every so often, for about 30 minutes until the rice is tender. You might need to add a little extra milk if it thickens too much at the end.

Alternative to pudding rice: japonica or risotto rice (the latter might need a little extra liquid)

Variations:

Clotted cream – use clotted cream instead of double cream for extra richness.

Cardamom, rose water and pistachio – swap half the milk for coconut milk, add a few drops of rose water and 4 cardamom pods instead of the vanilla and garnish with chopped toasted flaked almonds or pistachios.

Scandinavian risgrynsgröt – swap the vanilla for a cinnamon stick and serve with a cherry or berry compote.

Chocolate – add 100g chocolate towards the end of cooking in chunks and allow to melt into the rice.

Summer – make the rice pudding but add 100ml extra cream or milk at the end. Make a simple compote with 200g berries, such as blackcurrants, raspberries or blackberries, and let them cook gently with 2–3 tablespoonfuls of sugar until they collapse, then cool. Spoon the rice into 6–8 glasses and top with the compote. Serve cold from the fridge.

Healthier – use 850ml skimmed milk instead of whole milk, cream and water.

Serves 4–6

750ml whole milk
100ml double cream (or use 100ml extra milk)
100g short-grained rice, such as pudding rice
1½ tablespoons golden caster sugar
1 vanilla pod, scraped, or 1 teaspoon vanilla extract with seeds

Thai sticky rice with mango

In Thailand, where I was lucky enough to be taught this simple recipe, they call this pudding *Koa Niew Maoung*. It is also made in the Philippines, but there it is mixed with malty Milo chocolate powder, called *Champorado* and eaten for breakfast. You don't need to serve much as it is very creamy.

Place the rice, unrinsed, into a bowl with 150ml cold water and leave to soak for 1–4 hours.

Put the rice and soaking water into a pan with a further 150ml cold water and the salt, cover with a lid and bring to the boil. As soon as the water is boiling, remove the lid and keep boiling for 6–7 minutes or until the water has nearly all evaporated. Return the lid and steam-cook over a gentle heat for a further 6 minutes, then leave to rest while you prepare the coconut.

Heat the coconut cream and sugar gently over a low heat until the sugar has melted. Add to the cooked rice and stir together. Serve with plenty of sliced ripe mango.

Serves 4

100g Thai sticky rice or
   glutinous rice
¼ teaspoon salt
200ml coconut cream
1½ tablespoons palm sugar
   or dark soft brown sugar
1 ripe mango, sliced

Chocolate and vanilla semolina pudding

Semolina is used mainly for making pasta and couscous, but we should not forget the joys of semolina pudding, which is traditional in many countries of the world. As comforting as a wool blanket on a cold day, semolina pudding is a doddle to make: the ground wheat is cooked with milk and sugar until thickened. Try adding a sprinkle of cinnamon along with a little extra sugar instead of the chocolate and serving with stewed fruits or a grating of lemon and orange zest for a citrussy taste.

Put the semolina, sugar, chocolate, milk and vanilla in a pan and heat gently. Once simmering, cook for 4–6 minutes, stirring all the time, until the semolina has thickened and the chocolate melted. Leave for 5 minutes before serving.

Serves 4

50g semolina
1½ tablespoons sugar
150g milk chocolate
600ml whole milk
1 teaspoon vanilla extract
   or 1 vanilla pod, seeds
   scraped out

Raspberry brûlée rice pots

This is a wonderful dinner party pud and a refreshing change to crème brûlée but still with the favoured 'crunch' on top! You can obviously vary what goes in the bottom depending on the season – I'm a big fan of poached rhubarb and sweetened cooked blackcurrants, too. I've also included a more traditional baked rice pudding below.

Melt the butter in a saucepan and add the sugar. Stir over a medium heat for a minute or two, then tip in the rice and stir into the melted sugar mixture. Add the milk, cream, lemon zest, bay leaves and vanilla and bring up to a simmer. Simmer gently for 20 minutes, stirring every so often.

Preheat the oven to 140°C/120°C fan/gas mark 1.

Place a few raspberries in the bottom of each of 6–8 ramekins. Put the egg yolks into a large bowl and whisk lightly. Strain the milk from the rice into a jug and whisk into the eggs. Add the rice to the egg mixture, and stir together, then divide between the pots. Bake in a roasting tin half filled with boiling water for 25 minutes or until the brûlées are set but still have a slight wobble in their centres. Cool, then chill in the fridge.

When ready, preheat the grill to high, sprinkle the brûlées generously with a thick layer of sugar and place under the grill as near to the heat as you can (alternatively, use a blow torch if you have one). Grill until the sugar is melted and slightly browned, then leave to set to a crunch before serving.

Alternative to pudding rice: japonica rice

Variation: Traditional baked rice pudding (Serves 4–6)
Start the recipe in exactly the same way but omit the bay leaves. Add the simmered rice to just 1 whisked egg yolk and pour into a buttered 1.5 litre baking dish. Sprinkle with grated nutmeg and bake at the same temperature for 1½ –1¾ hours or until the rice is soft. (Serves 4–6 people.)

Serves 6–8

25g butter
75g golden caster sugar
100g pudding rice
850ml whole milk
200ml double cream
zest of 1 lemon
2 bay leaves
1 vanilla pod, split and the
    seeds scraped out
24 raspberries
4 free-range egg yolks
approximately 5 tablespoons
    granulated sugar, for the
    brûlée

Black rice pudding with coconut and pandan leaves

In the past, black rice was eaten less by locals and mainly by the rich and the wealthy. In Asia nowadays it is still not an everyday grain as it is more expensive to buy (because it takes longer to grow than other rice), but you can still buy it everywhere – not just in Asia, but worldwide. The short-grained black rice is used mainly for puddings and cakes. The traditional recipe for black rice pudding, taught to me in Bali, contains only palm sugar, but to Western palettes it has quite a strong caramel taste, which can be a little overpowering. This version is close to authentic but uses half light brown and half palm sugar. I have also added a scraped vanilla pod to enhance the flavour of the pandan leaves (known as Asian vanilla) – but feel free to alter to your tastes!

Soak the black rice for 24 hours in cold water, then drain and put into a large saucepan. Cover with boiling water and boil for 1 hour with the two pandan leaves and the salt, topping up with water as needed.

Add the washed white rice and vanilla to the black rice and continue to boil for 20–30 minutes, until both the rices are soft. There should still be a bit of water in the pan by the end of the cooking time – about a third full. Add the sugars and stir over the heat until the rice is starting to become a bit sticky, but don't let it catch the bottom of the pan – it should still be liquidy but thickened. Add the coconut milk and stir well. Serve with the fruit.

Alternative to black short-grain rice: japonica or pudding rice

Serves 8–10

250g black short-grain rice
2 pandan leaves (optional)
2 good pinches of salt
250g white short-grain rice, such as pudding rice, washed thoroughly twice
1 vanilla pod, seeds scraped and added with pod
50g palm sugar, chopped roughly (or use 50g light brown sugar instead)
50g soft light brown sugar
200ml coconut milk

To serve:
Jack fruit, durian fruit or sliced banana

Scottish honey and oat praline ice cream

Tasty and rich, this is a grown ups' ice cream that becomes even more grown-up if you add a little dash of Drambuie or whisky to the custard before churning. The idea for it is loosely based on the Scottish pudding cranachan, which contains all the elements – oats, honey, whisky and raspberries – stirred into whipped cream.

Heat the cream and milk together until very hot but not boiling, then stir in the vanilla seeds and pod and honey. Use a whisk to just break up the egg yolks. Remove the vanilla pod, then pour the warm milk and cream mixture over the egg yolks, whisking thoroughly as you go. Transfer back into the saucepan and cook over a low-medium heat, whisking all the time, for 6–8 minutes or until the mixture has thickened very slightly (do not let the mixture boil). Add the Drambuie or whisky, if using.

Pour into a new bowl and stand in a sink of very cold water to cool rapidly, then chill until cold. Pour the cooled mixture into an ice-cream maker and churn until nearly frozen; alternatively, pour into a shallow plastic container and freeze until three-quarters frozen, then thoroughly break up the ice crystals that have formed.

While the ice cream freezes, heat the oats in a dry frying pan for 3–4 minutes, stirring over a low heat until they are toasted but not burnt (it is very easy for the oats to catch and the result will be a bitter flavour so watch them carefully). Transfer to a bowl. Heat the sugar gently in a pan with 5 tablespoonfuls of water. When the sugar has dissolved, turn up the heat and, watching it carefully, boil until it turns a rich caramel colour. Immediately add the toasted oats and stir together, before transferring to a small tray or plate lined with greaseproof paper, to cool. When brittle, place the praline into a bag, seal and bash with a rolling pin until really broken up to an even crumble. Set aside until ready to add to the ice cream.

Put the raspberries in a pan with the jam and heat to collapse the fruit slightly. Strain through a sieve into a bowl and leave to cool.

When the ice cream is three-quarters set and you have stirred it (if not using a machine), add the praline and stir. Transfer half of the ice cream to a new container, then pour in half of the raspberry, then add the remaining half of ice cream and raspberry. Gently swirl to ripple but don't mix the raspberry in, before freezing completely. Remove from the freezer 10 minutes before serving.

Serves 4–6

600ml double or whipping cream
250ml whole milk
1 vanilla pod, seeds scraped out, or 2 teaspoons good-quality extract
150ml well-flavoured honey, warmed slightly if necessary to loosen
6 medium free-range egg yolks
125g raspberries
2 tablespoons raspberry jam

For the cranachan crunch:
50g rolled oats
5 tablespoons granulated sugar

'Wow your guests' easy soufflés

A great foolproof recipe for prepare-ahead soufflés, which literally won't let you down! Raspberries and blackcurrants can be used instead of the chocolate for a wonderful fruit alternative.

Butter six ramekins and coat with sugar (I add a bit of sugar to the base of the ramekin and then just turn it round in my hand until the sides are coated; then tip any remaining sugar into the next ramekin).

Heat the chocolate squares, 300ml of the milk and the vanilla gently, stirring, until hot and the chocolate has melted. Add the semolina and 1 tablespoon sugar and whisk for 3–4 minutes over the heat until thickened. Cool for 4–5 minutes, then beat in the remaining cold milk and the egg yolks. (At this stage, you can put the mixture and egg whites into separate bowls, covered, in the fridge for up to 4 hours, pressing some clingfilm onto the semolina so that it doesn't form a skin.)

When you are ready to cook the soufflés, take the semolina mixture out of the fridge and beat. Preheat the oven to 200°C/180°C fan/gas mark 6. Put a baking tray in the oven too.

Beat the egg whites in a large, clean mixing bowl using an electric whisk until they form stiff peaks, then sprinkle in the remaining 3 tablespoons sugar in about four batches to create meringue, whisking between each. Fold 1 table-spoonful of the meringue into the semolina mixture, then fold in the remaining meringue. Quickly spoon into the ramekins and fill right to the rim, levelling off with a knife. Run the tip of your thumb around the rim to clean around the edges so that the soufflés rise well.

Pop onto the baking tray and bake for about 10–13 minutes or until well-risen and brown on top. Serve straight from the oven onto plates and onto the table to wow your guests – they will sink if you hang around!

Variation: Raspberry soufflés
In place of the chocolate, use 300g fresh or frozen raspberries or mixed summer fruits such as blackcurrants, blackberries and redcurrants. Heat the fruit with 2 tablespoonfuls of caster sugar until the sugar dissolves and the fruit collapses then pass through a sieve to remove the pips. Beat into the cooked semolina before adding the egg yolks (you won't need to add the extra 50ml cold milk as per the chocolate recipe).

Serves 6

unsalted butter, for greasing
caster sugar, for coating
100g dark chocolate, broken
  into squares
350ml milk
1 vanilla pod, scraped, or
  1 teaspoon vanilla extract
40g semolina
4 tablespoons caster sugar
3 medium free-range
  eggs, separated

Cypriot lemon and semolina cake

Cypriot cooking is greatly influenced by the island's Mediterranean neighbours, where lemons, semolina and Greek yogurt are all commonly used ingredients. The semolina in this easy-to-make traybake adds wonderful texture and the yogurt gives an extra zingy flavour. It's a drizzle cake really, with the lemony syrup poured over the still-warm cake and then left to cool and turn deliciously sticky.

Grease and line a 20 x 25 x 5cm traybake tin. Preheat the oven to 170°C/150°C fan/gas mark 3.

In a bowl, cream the butter and sugar until pale and fluffy using an electric whisk. In a separate bowl, use the electric whisk to beat the eggs, yogurt, milk, lemon zest and juice until thoroughly combined.

Whisk the egg mixture into the butter and sugar in three batches, whisking well between each addition. Sprinkle over the semolina, then sieve over the flour and baking powder and fold in using a large metal spoon. Spoon into the baking tin and spread out lightly. Bake in the oven for 40 minutes or until firm but springy.

When the cake is nearly ready, place the syrup ingredients into a pan and boil for 10–12 minutes or until just thickened to a light syrup. Using a thin kebab stick, make 10–12 incisions in the cake while it's still warm and pour over the syrup (there will be quite a lot). Leave to cool, then cut into squares.

Makers 12–15 squares

175g unsalted butter
200g sugar
4 medium free-range
 eggs, beaten
4 tablespoons Greek yogurt
4 tablespoons milk
zest of 2 lemons, juice of
 ½ lemon
150g semolina
150g plain flour
2 teaspoons baking powder

For the syrup:
150ml water
175g golden sugar
juice of 1 lemon
1 cinnamon stick

Dutch apple tart with oatmeal streusel topping

This delicious streusel-topped recipe is actually what Americans call a Dutch apple tart; the traditional Dutch recipe is, in fact, topped with a pastry lattice. Streusel is a sort of crumble and the addition of rolled oats adds even more texture to it. A handful of sultanas is a lovely addition. I think a handful of summer berries would work well too, added to the filling to give it a summery twist. Serve the tart hot or warm with ice cream or softly whipped cream.

You will need a loose-bottomed metal tart tin with a base diameter of 24cm and a depth of 3cm.

For the pastry, cream the butter and sugar in a food processor, then add the flour and a pinch of salt and pulse about three times, scrape around the processor bowl and pulse again a couple of times. Add the egg yolks and process until it forms crumbs. Then add 2 tablespoons cold water and pulse until combined.

Turn the pastry out into your tart case and, working with cold hands and as quickly as you can, use your finger tips and knuckles to push the pastry into the case as evenly and thinly as possible. Pinch the pastry up the sides, building up the pastry so that you end up with the edges coming 6–7mm above the edges of the case. Push flat fingers into the base so that the pastry is as even as possible, though it doesn't have to be dead flat. Place in the freezer for about an hour (or longer, as you can blind bake directly from the freezer).

Preheat the oven to 200°C/180°C fan/gas mark 6. Scrunch up some grease-proof paper and spread over the pastry case. Fill with baking beans and place in the oven for 15 minutes. Remove the beans and paper, brush the base with egg and return to the oven for 5–8 minutes or until the base looks pale brown in colour. Then leave to cool.

For the filling, melt the butter and add the sliced apple and the three spices. Cook, stirring over a low-medium heat, for 10 minutes or until the apples start to soften. Add the sugar, the lemon juice and nearly all the lemon zest and stir over the heat to combine, then sprinkle over the flour and stir. Spoon into the pastry case, level it out and leave to cool slightly while you make the streusel.

For the streusel, in a bowl combine the flour, cinnamon, brown sugar, oats, a pinch of salt and the remaining lemon zest. Mix together, then rub in the butter until the mixture is crumbly. Sprinkle the streusel on top of the apples. Reduce the heat to 180°C /160°C fan/gas mark 4. Bake the tart for 30 minutes, or until the topping is browned and the apples are tender.

Serves 6–8

For the sweet pastry:
125g unsalted butter, cubed
75g icing sugar
250g plain flour
a pinch of salt
2 free-range egg yolks, plus a little egg for brushing

For the filling:
25g unsalted butter
1.2kg sweet and juicy apples, such as Royal Gala, peeled, cored and sliced
½ teaspoon ground cinnamon
¼ teaspoon ground nutmeg
¼ teaspoon ground allspice
75g caster sugar
zest of 1 lemon and juice of ½
2 tablespoons plain flour

For the streusel:
125g plain flour
½ teaspoon ground cinnamon
100g light muscovado brown sugar
100g rolled oats
1 teaspoon lemon zest
100g cold unsalted butter, cubed

Oats
Latin name: Avena sativa

Oats probably originate from a wild oat that grew in the Fertile Crescent of the Middle East. What are thought to be the oldest known oat grains were found in Egypt and date from around 2000BC, though these were probably weeds and not actually cultivated by the Egyptians. The earliest probable evidence of cultivation was found in caves in Switzerland and dates perhaps from the Bronze Age, relatively late compared with some other cereal grains. While its cultivation spread, throughout history oats were often disparaged as weeds, good enough only for animals. Scotland was one of the few places where the humble oat was appreciated, and it has been a staple of the Scottish diet since medieval times; oats are a hardy crop and grow well in the harsh northern climate. They were introduced to North America by the Pilgrim Fathers in the early 1600s but they didn't hit the commercial big time until the Ohio-based Quaker Mill Company started mass marketing them in the late 19th century. Quaker Oats are now a familiar brand around the world.

Oats are grown in most of the temperate regions of the world, especially in the United States, Canada, Russia and northern Europe. Though more than 80 percent of the global crop is still used as animal feed, oats are nowadays a common ingredient in Northern European cooking, particularly in Scandinavia and Scotland, where they are used in haggis, black and white puddings and dumplings, as well as porridge, of course.

Many different names are used for the various oat products on the market, and this can be confusing; the important thing is to look at the texture rather than the name. All oats start off as groats, and these are then cut, steamed and rolled or milled – the degree of processing determines the taste, texture and cooking time. You can buy whole, cleaned groats, but they need soaking and take a long time to cook. Porridge oats are grains that have been cut, steamed and rolled until they are as flat as possible; these small, fine oats absorb liquid quickly, making them perfect for porridge. Jumbo oats (also labelled oat flakes) are rolled whole groats, which keep their shape and are great for things like flapjack or muesli. Oatmeal comes in varying degrees of coarseness. The coarsest

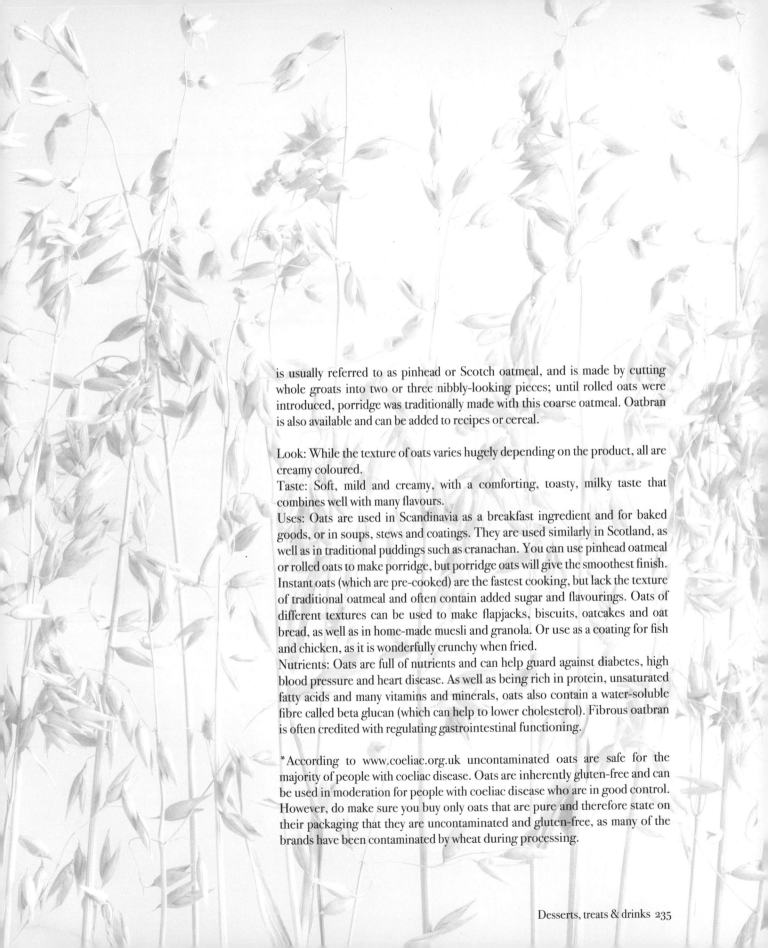

is usually referred to as pinhead or Scotch oatmeal, and is made by cutting whole groats into two or three nibbly-looking pieces; until rolled oats were introduced, porridge was traditionally made with this coarse oatmeal. Oatbran is also available and can be added to recipes or cereal.

Look: While the texture of oats varies hugely depending on the product, all are creamy coloured.

Taste: Soft, mild and creamy, with a comforting, toasty, milky taste that combines well with many flavours.

Uses: Oats are used in Scandinavia as a breakfast ingredient and for baked goods, or in soups, stews and coatings. They are used similarly in Scotland, as well as in traditional puddings such as cranachan. You can use pinhead oatmeal or rolled oats to make porridge, but porridge oats will give the smoothest finish. Instant oats (which are pre-cooked) are the fastest cooking, but lack the texture of traditional oatmeal and often contain added sugar and flavourings. Oats of different textures can be used to make flapjacks, biscuits, oatcakes and oat bread, as well as in home-made muesli and granola. Or use as a coating for fish and chicken, as it is wonderfully crunchy when fried.

Nutrients: Oats are full of nutrients and can help guard against diabetes, high blood pressure and heart disease. As well as being rich in protein, unsaturated fatty acids and many vitamins and minerals, oats also contain a water-soluble fibre called beta glucan (which can help to lower cholesterol). Fibrous oatbran is often credited with regulating gastrointestinal functioning.

*According to www.coeliac.org.uk uncontaminated oats are safe for the majority of people with coeliac disease. Oats are inherently gluten-free and can be used in moderation for people with coeliac disease who are in good control. However, do make sure you buy only oats that are pure and therefore state on their packaging that they are uncontaminated and gluten-free, as many of the brands have been contaminated by wheat during processing.

Chocolate-dipped Anzac biscuits

A slightly altered recipe from the original sweet oat biscuits which were sent overseas from Australia to the serving Anzacs in World War I. I reckon the troops would have been happy with the added almonds and chia seeds though, as they add nutrients and a delicious texture to these slightly chewy, treacly cookies. The chocolate is optional, but is a delicious addition if you have some in the cupboard.

Preheat the oven to 160°C/140°C fan/gas mark 3. Line two baking trays with greaseproof paper.

Put the butter and golden syrup into a saucepan and gently melt. Meanwhile, combine the flour, oats, coconut, almonds and chia seeds, if using, in a bowl. Stir the brown sugar into the butter, turn off the heat and add 2 tablespoonfuls of water and the bicarbonate of soda. Make a well in the flour mixture and pour in the melted butter mixture, then stir to combine.

Take small tablespoon-sized amounts and roll into 26 balls. Space them 6cm apart on the tray and then press down lightly about two-thirds with a fork to semi-flatten. Bake for 15–20 minutes or until brown. Remove from the oven, leave for 5 minutes, then transfer to a rack to cool.

Melt the chocolate in a bowl over barely simmering water, then dip the biscuits halfway into the chocolate and place them carefully on a baking tray lined with baking paper. Leave to dry in the fridge or somewhere cold.

———————

Alternative to oats: triticale

Makes 26

125g unsalted butter
2 tablespoons golden syrup
150g plain flour
100g quick-cook oats
50g desiccated coconut
2 heaped tablespoons
  flaked or slivered almonds
  (optional)
1½ tablespoons chia seeds
  (optional)
100g soft light brown sugar
1 teaspoon bicarbonate
  of soda

For dipping:
150g dark chocolate

Apricot, chia and coconut flapjacks

The mix of jumbo and smaller oats in these flapjacks is really for texture and means that you'll get a much better and less-likely-to-fall-apart end result! A combination of grains also adds extra nutrients. Triticale is a hybrid of rye and wheat, but in its rolled form it can be used in place of, or together with, jumbo rolled oats in muffins, bread or flapjacks.

Lightly grease a 21 x 25cm traybake tin.

Preheat the oven to 160°C/140°C fan/gas mark 3.

Combine the butter, sugar and syrup in a pan and heat gently until all are melted. Meanwhile, stir together the remaining ingredients in a large bowl. Pour in the melted sugar and butter mixture and combine together.

Pack the mixture into the tin and press down using your fists. Bake for 30–35 minutes or until light golden. Cool before cutting into squares.

Makes 16–20

125g unsalted butter
125g soft brown sugar
3 tablespoons golden syrup
125g jumbo oats or a mix of jumbo and rolled triticale
150g quick-cook porridge oats
100g dried ready-to-eat apricots, chopped
3 tablespoons desiccated coconut
1 tablespoon chia seeds (optional)
a pinch of salt

White chocolate crispies

I have very fond memories of making (and munching) crispy cakes during my childhood. A white chocolate version would have been a dream come true, but alas, I've only just realised how delicious they are! You could use raisins or nuts instead of the freeze-dried strawberries or blueberries, if you like.

Melt the chocolate in a bowl over barely simmering water, or carefully in the microwave. Stir in the puffed rice and strawberries or other fruit.

Either spoon the mixture into mini cupcake cases or into a 25.5cm x 23cm tin. Leave to cool for 2 hours in the fridge, then, if using the tin, cut into squares before serving.

Makes 20

175g good-quality white chocolate, such as Green & Black's
75g puffed rice
3 good handfuls of freeze-dried strawberries, crushed, or 50g dried blueberries, roughly crushed (optional)

Chocolate, cherry and hidden grain fridge bars

Served straight from the fridge, these are a favourite with young and old. I often cut them smaller and serve them after dinner with coffee. Ask your guests what the hidden ingredient is and I guarantee they won't be able to tell you. Adding raisins or frozen or fresh raspberries also works if it isn't the season for cherries.

Line a 21 x 25cm tin with a depth of 5cm with greaseproof paper.

First of all, cook the quinoa in boiling water for about 12 minutes, then drain thoroughly and cool.

Put the butter, cocoa, syrup and chocolate in a pan and melt over a gentle heat. Half fill another pan with some boiling water and gently heat. Break the topping chocolate into a heatproof bowl and leave to melt slowly over the water.

Meanwhile, break up the digestives into a large bowl and, using the end of a rolling pin, bash them up a bit more, so that they look roughly chopped. Stir in the quinoa, cherries, walnuts and chia seeds, if using. Pour over the butter-chocolate mixture, stir until combined and then spread out into the tin. Pour over the melted chocolate and chill for at least 4 hours until solid. Cut into 16–20 bars or squares and keep in the fridge.

Alternative grain to the cooked quinoa: cooked couscous or millet

Makes 16–20

75g quinoa
250g unsalted butter
3 tablespoons cocoa powder
5 tablespoons golden syrup
100g dark chocolate, roughly broken up
1 x 400g pack digestive biscuits
150g fresh cherries, pitted and chopped
75g walnuts, roughly chopped
1 tablespoon chia seeds, optional

For the topping:
150g dark chocolate

Berry shortbread bars

The addition of semolina is not original – many recipes for Scottish shortbreads include the grain as it adds a wonderful crunch. Brown sugar has a hint of caramel, which is why a combination has been used, but you can just use white sugar if you don't have brown in the cupboard.

Grease a 21 x 25cm traybake tin, preferably with a loose base. Preheat the oven to 180°C/160°C fan/gas mark 4.

Put the soft butter, sugars, vanilla and salt into a bowl and beat using an electric whisk or wooden spoon. Add the semolina and flour and whisk at a low speed until combined. Bring together using your hands, and then roll two thirds of the mixture out on a floured surface and press into the base of the tin. Put the remaining mixture in the fridge.

Bake for 20–25 minutes or until it is just beginning to turn light brown. Remove the tin from the oven. Mix together the raspberries and jam, squashing the fruit a little, then spread the mixture over the partially baked crust. Crumble the refrigerated shortbread dough over the jam to form a crumbled topping, then scatter with the almonds.

Return the tin to the oven and continue baking for 30 minutes or until golden. Remove from the oven and, when semi-cooled, carefully transfer to a wire rack and let cool completely. Cut into bars, then dust with icing sugar.

Makes approx. 16

150g unsalted butter, softened
50g soft light brown sugar
75g golden caster sugar
1 teaspoon good-quality vanilla extract
a good pinch of salt
100g semolina
200g plain flour
125g raspberries, fresh or defrosted from frozen
4 tablespoons berry jam, such as raspberry, blackberry or strawberry
50g flaked almonds
icing sugar, for dusting

Polenta and ricotta berry torte

Fine-ground Italian polenta, or ground cornmeal as it is also known, is a fabulous addition to sweet puddings and baking recipes, adding texture and soaking up all the wonderful flavours that you might choose to add. Serve this summer fruit torte either warm from the oven or cold from the fridge. It can be eaten as a cake or a pudding and needs only some cold cream and a few extra berries scattered around.

Preheat the oven to 160°C/140°C fan/gas mark 3. Grease and line a 23cm loose-bottomed, deep cake tin.

Cream the butter and sugar in a large mixing bowl until light and fluffy. Add the ricotta or yogurt, the lemon zest and 1 egg and whisk, then whisk in a second egg and then a third. Fold in the polenta, almonds and baking powder, followed by the berries.

Turn into the cake tin and smooth over, then bake on the middle shelf of the oven for 1 hour to 1 hour 10 minutes, or until a skewer comes out clean when inserted into the middle of the cake. Leave for 10 minutes before turning out of the tin. Top with the extra berries and a dusting of icing sugar and serve warm or cool with a dollop of cream.

Serves 6–8

175g soft butter
225g golden caster sugar
100g ricotta or natural
   yogurt
zest of 2 lemons
3 medium free-range eggs
175g fine-ground polenta
100g ground almonds
2 teaspoons baking powder
150g blueberries
150g raspberries

To serve:
100g raspberries
100g blueberries
icing sugar, for dusting
cream

Sticky salted caramel popcorn

Popcorn is by no means a modern invention. Evidence suggests that it was eaten in ancient America, and European colonists in the 16th and 17th centuries ate it with milk as a cereal. Its global reach as a popular food snack originated in the United States, however. During World War II, sugar rations prevented sweet production, so Americans began to eat popcorn instead. It has since spread around the world as the snack of choice to munch in cinemas. This salted caramel version takes 5 minutes to make and children adore listening to the popping noise, watching the transformation and, of course, eating the golden treasure at the end. This is especially true of my brother's daughter Lola who, when testing the recipe and trying the results, found to her delight that the entire bowl of popcorn lifted out in one piece! Thankfully, spreading the popcorn on the tray to dry has prevented that from happening again!

First, make the popcorn. Heat the oil in a lidded saucepan, add the corn and put on the lid. Heat, shaking the pan every so often, and wait for the popping to begin. Keep the lid on and shake until nearly all the corn has popped, then set to one side.

Put the sugar into a saucepan and gently heat with 1 tablespoonful of water until the sugar has melted (do not stir, just swirl the pan if you need to combine the mixture). Raise the heat, bring up to the boil and bubble for 2 minutes or until the syrup has turned a pale golden colour. Remove from the heat. Add the butter and salt (if using peanuts, add these to the popcorn just before you pour in the caramel) and beat for a minute or until the butter has combined with the syrup and thickened a little.

Pour the caramel over the popcorn and quickly stir together using two spoons until it is all coated. Spread onto a tray and break up the clumps, then cool for a few minutes until hardened and ready to serve.

Variation: Banana and peanut popcorn sundaes
Make the popcorn as above, but stir in a handful of roughly crushed peanuts or pecans to the popcorn, just before tipping in the caramel sauce. Scoop ice cream into bowls or glasses, layering with sliced banana and topping with a pile of the popcorn. Serve immediately.

Makes 4 generous cups of popcorn

1 tablespoon sunflower oil
4 tablespoons (50g) yellow popping corn
4 tablespoons (75g) caster sugar
40g unsalted butter
3 pinches of sea salt or 1 tablespoon salted peanuts, roughly crushed (optional)

Lemon barley water

We all know barley water, from the years it featured on Centre Court at Wimbledon as the favoured drink – or the favoured sponsor (I'm not sure which!). However, barley water has been around for far longer than tennis tournaments. There is evidence that it was drunk by the ancient Greeks, though the drink would not have contained fruit, perhaps just a little mint and honey. The recipe we make today combines pearl barley with water, sugar and fruit to make a squash or cordial. It's very easy, tastes unbelievably good and is much better for you than most of the additive-spiked squashes out there. You can make the barley water even prettier and tastier by adding some raspberry juice or crushed mint to the jug before serving. As it doesn't contain any added acid I would make it in smaller batches and just keep it in the fridge for a few days, rather than weeks.

Put the pearl barley into a pan with 700ml cold water and slowly bring up to the boil, then simmer for 20 minutes.

Meanwhile, peel the skin from the lemons as thinly as you can to avoid too much of the white pith and place in a large bowl with the sugar. Roll the lemons under your palms on a hard surface to squash them a little (this makes juicing easier), then squeeze and set the juice aside until later.

Pour the barley and hot cooking water over the sugar and peel and stir until dissolved. Let it stand and steep for an hour before adding the lemon juice. Stir, strain into a bottle and chill. Dilute with iced water or soda before serving.

Variations:
Citrus barley water – use 4 lemons and 2 oranges to make the cordial, using both zest and juice as in the original recipe.

Pink lemonade – add 1 part puréed and sieved raspberry (sweetened a little if the raspberries are tart) to 1 part lemon barley and 5 parts soda.

Vodka whizz – crush a handful of mint with ½ teaspoon caster sugar, then add 1 part lemon barley, with 1.5 parts vodka and 5 parts soda.

Makes 700ml undiluted
(serve 1:5 parts water)

125g pearl barley, rinsed
   and drained
5 unwaxed lemons
200g caster sugar

Index

Bibliography

BOOKS

A Cook's Guide to Grains by Jenny Muir, Conran Octopus, 2002
Mediterranean Grains and Greens by Paula Wolfert, Kyle Cathie, 1995
The Rice Book by Sri Owen, Frances Lincoln, 2003
The Cooks Encyclopedia by Tom Stobart, Grub Street, 2004
Plenty by Yotam Ottolenghi, Ebury Press, 2010
The Moro Cookbook by Samuel and Samantha Clark, Ebury Press, 2003
Classic Food of China by Yan-Kit So, Papermac, 1994

ONLINE

www.bbc.co.uk – www.celiac-disease.com – www.couscousdari.com – www.defra.gov.uk – www.edenfoods.com – www.freekehlicious.com – www.glutenfreeeasy.com – www.hgca.com – www.irri.org – www.italianfood.about.com – www.japanfocus.org – www.japan-guide.com - www.kamut.com – www.livestrong.com – www.myjewishlearning.com – www.nabim.org.uk – www.nytimes.com – www.riceassociation.org.uk – www.sciencemag.org – www.sharphampark.com – www.slowfoodfoundation.com – www.sunnylandmills.com – www.teffco.com – www.wholefoodsmarket.com – www.wildrice.org – www.wholegrainscouncil.org

Journal of Agricultural and Applied Economics 42,2 (May 2010): 337–355
spice.stanford.edu/docs/145 - Stanford University
www.resilience.org/stories/2006-12-28/ethics-biofuels
www.farmersguardian.com/home/business/business-news/consumers-reject-gm-food/35603.article
www.economist.com/node/5323362
library.thinkquest.org/TQ0312380/wheat.htm
www.thestar.com/living/food/article/1157487–farro-is-a-trendy-grain-that-goes-by-many-names
blog.foodnetwork.com/healthyeats/2009/02/24/meet-this-grain-wheat-berries/
www.worldwatch.org/global-grain-production-record-high-despite-extreme-climatic-events
www.sostanza.com.au/polenta_history_6.html
www.world-foodhistory.com/2011/02/cornmeal.html
www.hort.purdue.edu/newcrop/afcm/rye.html
www.hort.purdue.edu/newcrop/afcm/forage.html
www.agron.iastate.edu/courses/agron212/readings/oat_wheat_history.htm
bioweb.uwlax.edu/bio203/s2009/lanik_greg/history_of_barley.htm
www.biomedcentral.com/1471-2148/11/320

Acknowledgements

A massive thank you to my chief taster, chopper, washer-upper and general right hand lady – dearest Helen Paguntalan. Also to Fiona Philip – I couldn't have done it without you!

To my faithful friends and family and chief recipe testers/copy readers. Sorry for filling up your inbox's and thank you for ironing out any bumps along the way: Claire Broadley, Lucy Carlbom, Emily Cheetham, Mel Clegg, Harry Cox, Sophie Edmunds, Andrea Elles, Clare Evelyn, Jane Drysdale (mum!), Tula Goodwin, Kate Harris (as if I'd miss you out!), Caroline Hulbert, Hilary James, Kit James, Helene Keech, Hamish Laing, Renee Lomas, Tania MacCallum, Sophie Martin, Elle Micklewright, Nina Millar, Jodie O' Doherty, Clare Pannell, Pascale Pergant, Charlotte Petch, Karen Pilkington, Ros Plummer, Francesca Rathbone, Jen Reynolds, Lucy Ridgwell and Nancy, Marcia Ritchie, Judy Snell, Livs Syrett, Emily Taberner, Rebecca Taylor and Alex Urquhart.

Enormous thanks to those who donated recipes for the book:
Sam and Eddie Hart of Quo Vardis and Fino
Anna del Conte
Max's Mum
Aviva Walton
Becca Taylor
Anna Broome
Jane Drysdale
Emily Cheetham

Jonathan, Annie, Patrick and Liz – you have really made the book 'sing' with such beautiful pictures and design. Thank you all for the hard work you put in to make *Amazing Grains* the gorgeous book that it is.

To Tom Soper, I loved exploring the padi fields and markets with you in Indonesia. You take such beautiful photographs. Thank you for dropping everything to join me and letting us use the portrait shot.

To my agent Heather and Claire at HHB agency – a huge thank you for everything.

To Emily Hatchwell for casting your eagle eyes over the book and remaining so patient with all the emails back and forth.

And finally a joint gargantuan thank you to Kyle Cathie and my book editor Sophie Allen, as well as the Kyle Books team for your endless encouragement, patience and guidance. I hope you are as thrilled as I am.

About Anna Lynch
BA(Hons), DipION(Nutr), ATMS

Anna studied nutrition at the acclaimed Institute for Optimum Nutrition in London. She has run successful nutritional practices in the UK and Australia, and now lives and works on the northern beaches of Sydney.
www.annalynch.ntpages.com.au